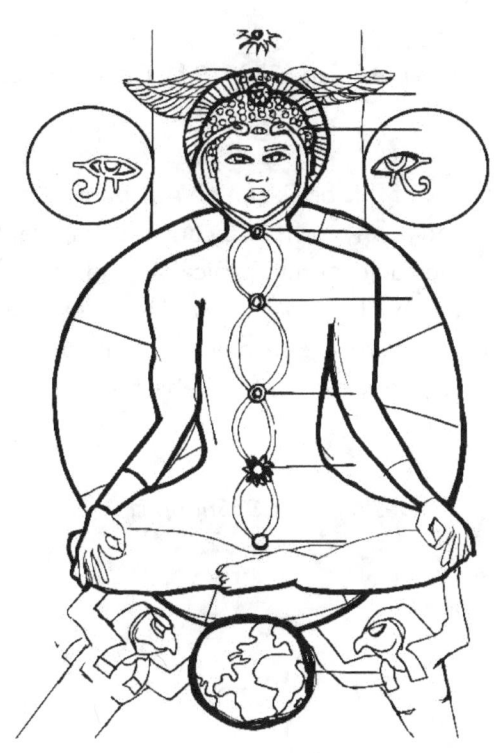

BLK Excellence:
Return to the Eternal Self

By L Ralliford

All Rights Reserved (C) 2019 by OoSnapp LLC

No part of this book can be reproduced or transmitted in any form or by any means, graphic, electronic, or mechanical, including photocopying, recording, taping, or by any information storage retrieval system, without permission of the publisher.

Designed by OoSnapp LLC

Author is Associated with the African American Writer's Alliance (AAWA)

You can also visit the writer's artist website: www.lralliford.us

For information address:

OoSnapp LLC

PO Box 12694

Seattle, WA 98111

www.oosnapp.com

IBSN: 9781695920347

This book is dedicated to people who find themselves feeling lost or trapped, in particular, it is for those who are constantly battling against being bullied. This book should inspire you to be aware of your true potential and value. Furthermore, I hope that this encourages you to keep pushing forward when the odds seem to be against you.

Contents of this Book...

The Purpose of this Book 8

Power and Humanity 18

Part I: Return to the Eternal Self

Reevaluation of Excellence 22

Disenfranchised Synthetics 30

More than One Dimensional 37

Dark Matter = Black Matter 43

Spirituality 51

Levels of Awareness 63

Introscoping for First Timers 71

The Importance of Identity 75

Sacrifice for the Ancestors 81

Libations 85

Creative Energy 91

Spiritual Evolution Theory 101

Cosmology 109

Tree of Life 143

Good ÷ Bad = Zero 149

Beginning the Conversation (Exercise I) 157

Part II: Power of Intent

Communication 169

Power of Thought 171

Thoughts are Louder than Words 179

Our Two Brains 189

Generational Challenges 201

Attack on the Millennials......................... 205

Symbolism 213

Chakra Points 225

Aura Layers 231

Basics of Numerology 235

History of Mathematics 249

Star Gazers 259

Knowing vs Believing 279

Spiritual Focus (Exercise II) 289

Part III: ZERO

Spiritual ZPE 293

The Tower and Fictional Karma 305

432Hz 317

Reiki Meditation Exercise (III) 335

Kemetic Focus Exercise (IV) 337

Part IV: Tarot Cards

Learning the Language 341

Major Arcana...................................... 347

Minor Arcana 373

Spreads 421

Final Response 427

Appendix 436

References 458

The Purpose of this Book

This book is for those who are lost and in search of some direction. For those whose nerves feel shot, and feel like they are a nervous wreck, in search of some substance. For those who are constantly battling against haters who surround them. For those, who also battle the haters that lurk deep inside of their minds. For those who doubt that they hold both power and value in this apathetic world. For those who fear to make mistakes. For those who struggle to find purpose. For those afraid of becoming "invisible" in this narcissistic digital superficial society.

This book is for you.

We are biological supercomputers, who are capable of building and altering our reality, however, we see fit. If we want to be surrounded by prosperity and happiness. All we ever need to do is set up the perimeters that will alter our reality, and BAMN... the world as we once saw it, has changed. We are products of change, that is constantly changing. We were never designed to remain stagnant. We are destined to strive to be better than we were before. When life seems to constantly be getting you down, it is hard to even imagine how you can alter your reality.

"Some people are just unlucky." I heard that statement from someone once. I was also told by my former mother-in-law, that *"One day Latoya, you will learn that some people aren't worth a damn thing. You best just learn to recognize those*

types of people because they will never produce anything good. They will only bring you down."

I try to remain optimistic. I want to believe that every person in the world can alter their reality, even the die-hard toxic people. If a man was a crook all his life, is he not capable of changing his ways? If a woman was addicted to drugs for most of her life, is she not capable of quitting and living a starting again? We each hold the keys to altering our reality. For example, let us look at the Triplets. The Triplets is a dance group that consists of children who are from the ghettos of Uganda. These children were on the brink of starvation and had ripped and worn-out clothes. Instead of accepting their reality, they decided to, well, dance on camera. Their dance moves went viral and made them famous. As a result, this completely changed their lives, as well as other misfortunate children living in similar situations. In that single moment when they decided to spontaneous change the mood, they unconsciously altered their reality. Something clicked inside of them, and their energy shifted. This shift in energy quickly inspired a wave of others to create a movement of "dancing in the ghetto" type videos off throughout the world. These "dancing in the ghetto" type videos also inspired celebrities to showcase the social irony, in their creative works, such as the song: This Is America by Childish Gambino. In this music video, the black children danced while they were surrounded by extreme chaos. But while they danced, they appear to be happy. The energy surrounding the Triplets shifted and it also inspired others, and this produced a form of change. Change is an interesting thing. It may not be completely good or completely bad, but it is what it is. Like a pebble dropping into a pond. Change forms ripples that spread out and affect all around us. Thus, reality has been altered. But has their characteristic also been changed?

In some ways, yes it has. When the reality is altered, the character experiencing these changes are also forced to change, too. Even the hardcore toxic people, that my former mother-in-law said they would never produce anything. Everyone is universally forced to be altered when the reality has shifted.

The consequence of this universal shift is that willfully ignorant people are determined to never change (which is an action that goes against cosmic law). People who are determined to not be changed, are filled with negative energies (also known as demons) that twist their minds and disables their critical thinking abilities. These same individuals are willfully ignorant in some aspects of their lives, but they are also ignorant in other aspects. I do believe that the area that they are ignorant in, is the understanding of cosmic law and human nature.

Our lives go through a series of different mental stages, which continuously cycles. Our ancestors taught about these different stages of human awareness and human development, on different mediums. Some taught about this progression in stories. Others embedded this information into their creative works and artifacts. Some shared the information through music. And others created medium devices.

Medium devices like Tarot cards were created to teach us about the evolution of awareness. The Tarot Decks that we know now, were created in the medieval period (15th Century), in Europe. The decks were first used to play trick-taking card games in France, Italy, and Austria. It was later, in the 18th century, when the tarot cards became more popular to be used as divination practices. Though the creation of the Tarot cards is new, the characteristics associated with the cards are very old. The Major Arcana cards represent twenty-two different characteristics. These twenty-two characteristics are very similar to characteristics found throughout the ages in various religious deities. Additionally, the progression of the main story of the cards matches the progression of the Kemetic Tree of Life, in which specific Neteru (the general name or species of deities) represents the characteristics that we must imitate to ascend.

The more you study, the more you will notice that throughout human history people will unconsciously do things that are in line with cosmic law. The tarots are a great example of how a card game inevitably ended up being the perfect divination device that shares the same story of our ancestors or in line with the cosmic order. This leads us back to the story of the Triplets:

When I examined the success story of the Triplets, I discovered that their story closely resembles the characteristics of the Tower (a card in the Major Arcana).

The Tower is number sixteen in the Major Arcana set. Its numerical placement is also important to understand as it outlines a point in the protagonist's story when they are on the verge of facing a climax in this phase of their story. In the protagonist's story, this climax is filled with new problems that make the situation chaotic. There are different levels of chaos in our lives. Some chaotic experiences are more extreme than others. This is why in the Major Arcana, there are a few cards that can also symbolize some form of "chaos", like the Devil and Death. But the Tower card is the most extreme version of anarchy. A key thing to note is that for every situation, there is a resolution or advice. Hence even in a period of seemingly absolute anarchy, there is a key or clue that can resolve the situation.

In the case of The Tower, the clue to fixing the situation lies in the protagonist who is seeking someone else to come save them. It's a riddle and a story that has been retold over and over again. Every Hollywood movie follows the same story as the Tarots, in which the protagonist begins the story as The Fool. At some point, the hero will be in a chaotic situation and seeks someone else to come and save them. But in the end, they realize that they held the power, all along. The hero takes their newfound power and uses it to save the "world" or to remedy the chaotic situation. Then at the end of the story, the hero reaches a new level of awareness of themselves and the world around them – a fresh new terrain for them to explore, like The Fool.

The success story of the Triplets was just like that... They started unknowing of the depth of their situation. And amid hopelessness, they decided to change their perspective and dance. At that moment, they uncovered a new power that they did not realize they had, which changed the reality around them. And I imagine that at this point, they are seeing the world from a new perspective and beginning a new level of the journey.

The most real thing that you must accept is that: We are Products of Change.

Therefore, to go against this simple reality is to go against human nature. It is one thing for the world around you to be chaotic, but it is a different thing entirely for you to be internally chaotic.

***

At the time of writing this, a news report about a TSA Agent killing himself from jumping off a balcony has made all the headlines. Seeing that the report got me thinking about if only he had known how to reset his unstable energy, then perhaps the desire for suicide could have been diminished.

If only...

I write this to you because I am sure that some of you are experiencing similar feelings as that TSA

Agent, to where you feel as though you are trapped in a corner and have no other options. For those who feel this way, I feel compelled to connect to you, to let you know that you do not have to torture yourself by trying to keep up with the Jones.

In this digital age, information and situations are being thrown at us so quickly, that it is really hard to complete a full circle of thought before it is disrupted by some distraction. This constant disruption causes our internal energy to become distorted or chaotic. When we are distorted, it is hard for us to critically think through our situations. Perhaps with a clear mind, the TSA Agent (who probably killed himself due to the long-term government shutdown) would have seen a new solution to his problem. Such as looking for a new job. Or going to stay with family. Or working through the situation, one step at a time. Sometimes all it takes is for us to slow down, for us to find the solution we have been looking for.

With that said, this book is aimed towards teaching you how to slow down as well as to reconnect to your Eternal Self. In addition to learning how to slow down, this book teaches about the principles and conceptualized ideas behind African Spirituality.

Some are not like the TSA Agent, at all. These individuals are frustrated and angry and filled with rage, as well as fear. These individuals can be commonly identified as those associated with radicalism, from the nationalists, racists, bigots, fascists, purists, and so on. These individuals who practice radicalism are not easily identified by color or organization. A liberal can easily be as radical as a conservative. A vegan can be as radical as a meat-eater. A round-earther can be as radical as a flat-earther. A black man can easily be as radical as a white man, and both can be racists.

Radicalism is when an individual practice anything on the extremities. When a person lives on the extremity of something, they are opening themselves up to be vessels of an abundance of negative energies – or legions of demons. Moderation and balance are the only way we can reach a God-like state. This book is structured to help you to understand the different aspects of you and to teach you to live in balance. This book is divided into four parts to help you learn the full picture of internal power.

These parts are:

1. **Return to the Eternal Self:** exploring what the eternal self is.
2. **Power of Intent**: exploring how to identify intent and the unknown power you have to alter your reality or to reach other realities.
3. **Zero**: exploring how to reach a God-like state of mind.
4. **Major and Minor Arcana**: exploring the definitions of various spiritual characteristics and other names for the same entities. As well as learning about how to take control of the situation.

5. ***You are probably wondering what this has to do with blackness. But as you will learn in this book, we are all black including the Divine energy. Connecting to your truest self is to connect to truest blackness.***

Throughout this book are various interactive activities. These exercises are designed to help you develop spirituality. Additionally, you will notice that a lot of the content in this book is grounded in a lot of scientific referencing. As an author, I felt this is important to show you the different variations of the same concepts to help you connect. I have actively made a point to provide massive references and I suggest that you explore each reference so that you can see the full picture of why I referenced to those sources in this book. With that said, the appendix section of this book is very extensive. There are various facts and descriptions that I felt was also important to share that didn't necessarily have a place within the highlighted four sections.

The BLK Excellence Tarot Deck is designed to complement this book. Each card contains beautiful illustrations that combine a variety of different African cultures, traits, symbols, figures, history, and practices. At the top of each card are Medtu Neter numerals, making this a meaningful way to learn how to count in Kemetic. Additionally, these cards are embedded with various symbolism traditionally used in tarot cards. Which therefore combines African legacy with Spiritual Science and feels like they are a nervous wreck, in search of some substance. For those who are constantly battling against haters who surround them. For those, who also battle the haters that lurk deep inside of their minds. For those who doubt that they hold both power and value in this apathetic world. For those who fear to make mistakes. For those who struggle to find purpose. For those afraid of becoming "invisible" in this narcissistic digital superficial society.

This **book** is for you.

Power and Humanity

Look at the world around us.

It's magnificent in its glory.

The natural world is a part of...

Our misguided story.

Look at the world around us.

Look at the flowers;

Look at the birds...

Witness these natural powers

And be reassured...

That life is endless...

Love is embedded,

Within all of this.

It never ceases to amaze me...

That so many of us cannot see...

How love is pre-programmed

In you and me.

Love is embedded,

Within all of this.

Therefore, stop fearing to be

Touched or kissed.

Embrace these feelings,

We have been given.

Love is the savior...

Of life re-risen.

By savior, I do not mean,

White Christ.

By savior, I am referring to

Our inner-black light.

Power,

Lies boldly within.

You and I are immortal,

It is only our flesh that

Is dying.

We are the next generation of ancestors.

Every action we make

Every word we speak

Every ambition we take

Every struggle we defeat

Has been written

On the ribbon of time.

We are the collective

Divine.

Who is in constant battle within ourselves

We are the kings and queens

Destined to lead new generations

To better days,

Could you imagine a world,

Passed this faze?

Where our children's children

Are free to gaze…

And appreciate the beauty of life.

They could be free

To enjoy sunsets and sunrises

Free,

To become humanity's wisest

Free,

To discover the

furthest depths of Love

Free,

To explore the furthest reaches of space, above.

All the places

We want to get to

Begins with simply

Me and you.

It is time for us to think

Like Immortals…

And time for us

To surpass every hurdle

Let us discover

What we are capable of…

With Love.

Part I: Return to the Eternal Self

Reevaluation of Excellence

EX · CEL · LENCE (noun)

The quality of being outstanding or extremely good.

For more than 400 years, we have been conditioned into believing that the color and term "black" is primitive or the official representation of the lowest form of creation. This global assumption of "black being lesser" has conditioned our civilization into practicing the belief that "white is right" and that "white is good", even when blatant evidence contradicts this attitude.

We are living in an interesting time when the millennials are divided between either embracing or shunning this conformity. In this second millennium of the "common era (also known as CE), many of us have noticed the contradictions within the "white is right" notion and have discovered that in most cases, "white" is the product of black. As a product of black, in many of these situations, the fairer version of a previously darker substance, is less concentrated. This awareness that black is the stronger version of white, can be uncovered throughout colonialist history. Particularly, in the blatant practices of trying to create more docile African slaves, by mixing them with European DNA.

> *"It's just like when you've got some coffee that's too black, which means it's too strong. What do you do? You integrate it with cream, you make it weak. But if you pour too much cream in it, you won't even know you ever had coffee. It used to be hot, it becomes cool. It used to be strong, it becomes weak. It used to wake you up, now it puts you to sleep."*
>
> – Malcolm X

The society we reside in has convinced many of us to fear black and any form of blackness. Evidence of this programming to be *fearful* of blackness can be detected in various cultural practices. For example, black animals (dogs, cats, rabbits, etc.) are less likely to be adopted because of their color. Darker-skinned people are commonly feared and treated badly by other races of fairer complexion. Nations of people around the world fear to be "too dark" and bleach their skin, in hopes of looking more appealing. Or the fear to become "too dark" and avoid sunlight as much as possible, to avoid being tanned. Naturally, black products, like Vanilla, are bleached to seem more appealing to general consumers.

CNN news hired a team of psychologists and conducted a research report about colorism in the school system. In this report, eight public schools participated (4 in greater New York and 4 in Georgia). The reporters leading this report were Anderson Cooper and Soledad O'Brien, who did a report on children and how they have racial biases. Both black and white children were shown a series of images and dolls that's the only variation was in color.

> "A white child looks at a picture of a black child and says she's bad because she's black. A black child says a white child is ugly because he's white. A white child says a black child is dumb because she has dark skin"
>
> – Cooper (2010)

The results of this report shocked many viewers of this experiment. After all, how could our innocent children be racist? When you examine our culture (worldwide) you will realize that it is inevitable that our children would start to view color in this way. Culturally everything associated with the term white is the opposite of everything associated with the term black. This behavior is even embedded in modern language. It is to the point where the "opposite" doesn't even need to be a reference, the listener will automatically assume the existence of the opposite. Notably, the term black ice is commonly used in weather reports. In these same reports, there is no mention of white ice, even though white ice exists. If culturally you are expected to view everything associated with white as being docile, it is only natural that the instinctive reaction towards everything associated with black would be treating the black subject as dangerous. It is important to realize that ice is ice and regardless of its appearance, the actual substance is dangerous on the roads. It is delusional to view either version of this substance as lesser.

Meanwhile, as we further examine this concept of blackness, our opinion and bias about this term will begin to change. Outside of the conformities of modern society, we have the potential to discover that many substances, which would be categorized as black or dark, are rich in nutrients. We will also discover that the foundations of all creation are built upon darkness.

According to NASA in their article titled Dark Energy, Dark Matter:

> "What Is Dark Energy? More is unknown than is known. We know how much dark energy there is because we know how it affects the universe's expansion. Other than that, it is a complete mystery. But it is an important mystery. It turns out that roughly 68% of the universe is dark energy. Dark matter makes up about 27%. The rest - everything on Earth, everything ever observed with all of our instruments, all normal matter - adds up to less than 5% of the universe. Come to think of it, maybe it shouldn't be called normal matter at all since it is such a small fraction of the universe... First, it is dark, meaning that it is not in the form of stars and planets that we see. Observations show that there is far too little visible matter in the universe to make up the

27% required by the observations. Second, it is not in the form of dark clouds of normal matter, matter made up of particles called baryons. We know this because we would be able to detect baryonic clouds by their absorption of radiation passing through them. Third, dark matter is not antimatter, because we do not see the unique gamma rays that are produced when antimatter annihilates with matter. Finally, we can rule out large galaxy-sized black holes based on how many gravitational lenses we see. High concentrations of matter bend light passing near them from objects further away, but we do not see enough lensing events to suggest that such objects make up the required 25% dark matter contribution." (NASA, u.d.)

What makes this discovery even more fascinating, is that the Kemetic people documented a similar concept of both dark energy and dark matter, over 10,000 years ago.

Is it not exciting to know that we are all a product of blackness? This type of blackness goes well beyond skin color. This type of blackness is so black, that we are incapable of distinguishing the richness of its color. The foundations of the universe, including all of the particles and atoms, are formed from this matter. This is the energy that grounds us and fills us up with power and purpose. We see skin color on this physical plane or physical realm called Aye. But in higher dimensions of awareness, we are all black, as we have no physical form.

Black Lives Matter.

The more you are in tune with yourself and the universe, the more you will be able to identify interesting traits of the physical reality. In this dimension, people are bound to instinctively imitate actions of higher dimensions. That is what the statement "as above, so below" means. Everything that occurs above us in the heavens (or higher dimensions) will occur in this reality as well. The only difference is that unlike those who are aware in higher dimensions, those who are disconnected in this dimension are unconscious of their actions.

The popular statement "black lives matter" has always been puzzling to me. From a writer's perspective, I always viewed the statement as an incomplete sentence. There is no direction or conclusion or explanation for the statement. It is simply left open-ended. But when I began studying BLK Excellence, I began to see the true meaning behind the statement that most are unconsciously aware of. English is an interesting language, for starters. The word "lives" can be pronounced in two different ways to produce two different meanings.

In one sense, "lives" is a verb that can refer to multiple "lives", as in to remain alive. Or it can refer to living in a particular place. Thus, I translate the *"Black Lives Matter"* statement to read:

Black lives [in] **Matter** [and thus] **Black lives matter**.

As in everything, *the unconscious unconsciously imitates cosmic order.* This truth can be found in the statement, Black Lives Matter. White American bigots produced a counterstatement "All Lives Matter" as a means to shut down the Black Lives Matter movement. The paradox is that Black lives in Matter and thus Black lives matter, and since the matter is in everything, then it is true that "All Lives Matter" as it is true that all lives are essentially based on blackness.

Black Lives Matter = All Lives Matter

In other words, even though the bigots attempted to diminish the Black Lives Matter statement, the same bigots were unintentionally saying the same thing or unintentionally agreeing in the essential value of blackness. This realization leads us to another important factor, which is self-hate. If a man hates everything black and his entire being is based on blackness, then that man is not simply a racist but a victim of self-infliction and self-hate.

It is the most real and oldest form of justice... the reality of identifying that we are all the one color. Consequently, you should start to understand the concept of why black is excellent. As it exists in everything and everything is the product of black.

Our ancestors understood the connection between the spiritual world and our reality and the power of blackness. Their knowledge was vaster than what we can currently comprehend. However, even from the mere fragments that we can decode, we can witness pure greatness. The energy that created us and the power that is within us, is black energy and our spirits, our Eternal Selves, have a seat amongst this black energy... our Eternal Selves are sitting within the Divine Black Energy. **Therefore, the act of practicing BLK Excellence is practicing how to reconnect with our Eternal Selves or practicing how to reconnect to true blackness.** By reconnecting to this power, we can open new doors and find new opportunities that will help enhance our lives.

Disenfranchised Synthetics

The original structure of existence is built from the darkness. As mentioned earlier, our society is programmed to believe that righteousness is white. Therefore, to maintain this perception (whether to unconsciously keep up with demand or tactfully created to maintain this prejudice) many natural things are bleached white to be more appealing. Modern society seems to surround itself with synthetics. The current trend seems to be that everything natural, is wrong. This concept is far from correct.

This mania over synthetics has pushed many people into being disenfranchised with their selves and their roots. It is important to know that a plant with no roots will quickly wither and die. If you do not love yourself for what it is, then you will struggle to know what it is that you want out of life or know what direction to take.

> "Some people black and ugly. Some girls black and ugly, you know what I am saying. So, you see when they put on bleachings, they kind of look good… Yeah because there are some girls on the road that are too Ugly. Their complexion is black like a tar…" said one commentator from the documentary called Skin Bleaching in Jamaica (JoyDailyTV, 2016)

Jamaica is just one of many places in this world that has become obsessed with bleaching. When individuals have been asked why they are obsessed with bleaching, many of them state that they want to be like Michael Jackson. Regardless of the horrible damage that they put on themselves, many of these individuals believe that it is still better to continue to bleach themselves. Michael Jackson, by the way, suffered from serious health problems and was in constant pain. It is speculated that Michael Jackson was suffering from skin cancer, which might have been associated with his drastic skin alteration. In the Telegraph's article Michael Jackson's History of Health Problems, author Heidi Blake states that:

"In May this year, the singer's [Michael Jackson's] 50 comeback concerts at the O2 arena were rescheduled amid reports he was battling with skin cancer, which his publicists denied. Randy Phillips, President, and CEO of AEG Live, the show's producers, said the schedule changes were needed to provide more time to manufacture the set and prepare for dress rehearsals. The publicist dismissed the health fears, saying 'He's not dying of skin cancer.' But insurers were said to be reluctant to cover the string of London concerts, forcing AIG to take on some of the risk itself. Although organizers say Jackson passed a full medical before the concerts were scheduled, he was frail enough to require a wheelchair..." (Blake, 2009)

Regardless of the obvious negative side effects of the practice of bleaching, the craze of altering one's appearance in this way has been increasing at an alarming rate. It is not just skin bleaching that is a trend. Plastic surgery and body implant practices have been on the rise as well. It makes us wonder if this need to alter the original self reflects universal evolution or is this act and act of human rebellion against nature.

I believe that it is a combination of both and inevitable, which is the nature to rebel as well as nature to imitate the universe. Before explaining the reason behind that conclusion, it is first important to point out that in this society we are obsessed with primitive and sexual energies.

Primitive Energy is the instinctual persona that we all have embedded in us. When humans were prehistoric this form of instinct was the fight or flight notion, that our predecessors used to survive. This energy is stored in the base of our spine, which is most commonly known as the Root Chakra. According to Dr. Freshwater: "The vital life force energy of Root Chakra begins at conception and continues with life force energy dividing the cells during Embryogenesis. The spinal column begins to develop from the bottom (root chakra) moving to up (crown chakra), and then organs begin to develop." (Shawna Freshwater, Ph.D., 2017)

Above the Root Chakra point is the Sacral Chakra point, which is the source of our sexual energy and both of these energy points reside on Aye (the physical realm of our reality). These points are considered to be the "lower half" of the human body, which relies and operates primarily on primitive instincts and takes instructions primarily from our second brains. As it stands, our culture is captivated by the autonomy of the lower half of the body. There has been a massive push of butt implants that goes beyond social sanity. This movement to be more sexually appealing on the lower half (from the stomach down to the groin) seems to reflect overly stimulated chakra points. The process to stimulate the logical half of our bodies (upper half) is less interesting than focusing our attention on our lower half. Additionally, the strong need to "fix" our outward appearance, is more interesting than taking care of our internal wellness. This desire to focus on shallow aspirations is a sign of a society who has lost touch with their inner selves.

The side-effects of this trend of imitating someone else will forever cause individuals to become disenfranchised from true power. Many poor souls feel that the only way they can be acknowledged is by altering themselves and their lifestyles to match the current trend. Something important to take note of is that trends are very temporary and constantly changing. Today, it is cool to be brown but twenty years ago it was cool to be mixed. Today, butts are the most desirable feature, but twenty years ago it was facial features. Today, skinny jeans are cool but twenty years ago, pants needed to be baggy. Today, tattoos are cool but twenty years ago, tattoos were associated with the criminal lifestyle, and you were denied jobs for having them. The sad reality is that people will permanently disfigure themselves to meet the status quo of a trend that is designed to spontaneously change tomorrow.

Ironically enough the highest-paid models today were taunted for being too black, too fat, too tall, too short, too Africa, too pale and too patchy, over ten years ago. What makes these models so dynamic is regardless of what was trending in society, at the time, they chose to stay true to themselves. And thus, their reality conformed to them. Not them attempting to conform to that temporary reality.

Hence, if you still like rocking baggy pants but people tell you to wear skinny jeans. Keep on doing you, boo. I guarantee that the people who deserve to be named your friend, will not taunt you for not conforming to trends. People that deserve your love and respect, only seek your happiness and that you are content with who you are.

How do you obtain power? You obtain power by becoming spiritually connected and by embracing who you are. With that said, it is time to start understanding all the different aspects that makeup you.

More than One Dimensional

Firstly, we operate in multiple dimensions. Our bodies are the physical versions of the *Tree of Life*. In this concept of the Tree of Life, we operate on the highest dimension as well as the lowest dimensions found in the underworld. We are also able to continuously adapt. One of our biggest weaknesses is that we remain stuck in singular concepts or narrow-minded way of thinking. We are bound into this illusion of hypocritical behavior, where many quickly place labels on everybody that they deem to be a hypocrite. While upon further evaluation the individuals that are marked as hypocrites, typically are not being hypocritical. In contrast, those that are hypocrites are not marked as hypocrites but labeled as leaders that many aspire to be.

For example:

If 10 years ago, I told you that I did not like the taste of beer because I had found it to have a bitter taste. But over time, I explored the different flavors of beer and I found that I quite liked a summer beer and ale, from time to time.

This change of mind does not demonstrate hypocrisy but instead, it demonstrates merely a change of mind. People have the right to free will and the right to change our opinions. This is one of the factors of humanity that makes human beings' complex creatures. But many of us base our reality on artificial fantasy which is presented to us through the media. The media displays people (celebrities and characters) as being single dimensional beings. The anchor must *always* be happy. Or this protagonist's character must *always* be sad. While the villain of the story must *always* be mad. It is a comical tragedy that people are unconsciously imitating fictional character behavior that they were exposed to from their childhood shows in the 1980s until Today. Where in cartoons like the Care Bears and the Snorks, all of the fictional characters were lived their life being a single emotion. In all of these popular children's shows, the civilization of these fictional worlds depicted a population of one-dimensional beings who were named either after their powers or their singular personality.

The reason why there are so many superficial people may be due to the side effects of this style of mental programming. As anything that doesn't fit into a preset social box is labeled weird or an outcast or hypocritical. Throughout history, there are so many greats that have said one thing and then later said something else that contradicts their original statement. Regardless of this shift, their followers tend to only focus on one aspect of their story and dismissed the rest of their identity for the sake of social validation. This selective acknowledgment of individual accomplishments has become the fundamental bases trying to find answers for very complex questions, like:

Are the actions or statements of a man forever held against him?

Are the actions or statements of a man's ancestors forever held against him and his predecessors?

If an ignorant white man decided to call a black man the N-word, does that automatically make the white man a racist or does it mean that he is ignorant?

If a man (or woman) was the child of a pedophile, does that mean that by default he will also become a pedophile?

Are the descendants of slaves fated to be slaves by default?

Is an entire race permanently labeled as racist because their ancestors created racism?

Is maintaining these social double standards constructive or destructive?

There is a meditation activity which is cleansing each chakra point so that the 3rd Eye can open, and we can spiritually ascend. When cleansing your chakra point for the 3rd Eye chakra point, you must remove any illusions about yourself and your reality. Therefore, would not be that "double standards" are blocking an individual from ascending? In short, we are products of Change. We are expected to shift and move like a river. It is in the process of trying to stay the same, where we are faced with internal conflicts.

I have seen racism, and this experience was not found in an idiot throwing about a dumb insult because he is losing an argument. The racism I saw was pure hatred from these lost souls. It was so intense (this energy of fear was massive and hatred without cause) that I dare not approach them. I could sense it a few yards away... if not further. It was an energy that I dare not want to be around again. Not out of fear, but out of pity. After witnessing these poor souls trapped in this toxic reality, I found my heart was racing from the sense of their anxiety and anguish. As a psychic, I read people's spirits before I speak to them. I saw these trapped pitiful weak spirits suffocating from the toxic energy surrounding them. Will I run-up to one of this pitiful soul to give them hugs as a means to save them from their hatred? No.

That's not my forte.

I am not the one who can change them. They have to be the ones seeking the change. Luckily, many of them do. Lots of former white nationalists escape from these Hate organizations in a similar means as a victim escaping their domestic abuser. Once free of the toxic environment, they find love and happiness. These changed individuals are proof that a person is capable of changing their identities and morals when they have the desire too.

When I was writing for KERNE Magazine, I wrote an article about a mixed-raced man who was raised as a white nationalist by his white mother. His story fascinated me because, regardless of seeing his darker complexion in the mirror, he still hated everything that wasn't white. And then, he told me, something occurred to him in his late 20s and he realized how wrong he was. As a result, with all of his might, he sought out to learn more about his "black side" and became an activist for equity and Black Awareness.

His story inspires me because it reminds me of the power of individual will and that while we live, we have endless possibilities to adjust and tweak and repair our ideology. I think back to all of the celebrities who have been marked hypocrites because they did not fall in line with this specific "social stereotype" for a specific movement. If a black person marries a white person, does that mean that they have no right to categorize themselves as black and to stand up for situations related to their democratic? Can only "pure black" people speak on behalf of black people? Are mixed raced people ineligible to make remarks or speak on issues related to both races that they identify with?

We are not Care Bears. We weren't born with a symbol on our bellies which represents our one-dimensional character. We are given an arrange of emotions and feelings that help us experience the world around us. None of us were born with a manual programmed into us, about how we are meant to live. We are born to live a life and to explore the successes and failures of doing so. We should, therefore, be socially allowed to evolve... like the universe.

Dark Matter = Black Matter

As mentioned earlier, the matter (or particles) that we consider to be "normal" make up a mere fraction (about 5%) of the universe. Outside of our social norm, the entire universe is built upon the foundation of **Dark Energy** and **Dark Matter**. Dark Matter is defined as a hypothetical form of matter, which makes up about 25% of the universe. This type of matter gets its name as it is invisible to the electromagnetic spectrum and doesn't react to electromagnetic radiation (Light Particles). In astrophysical studies, the gravitational effects in dark matter remain a phenomenon. Dark Matter, according to various astrophysics and evolutionary theories, remains the key ingredient to the creation of the universe and life. Dark Matter surrounds normal particles (galaxies) like a dark halo. The Dark particles are constantly moving through the universe and our solar system and even us.

> "Every second, you'll experience about 2.5 x kilograms of dark matter passing through your body... Dark Matter, to the best of our knowledge, is out there in all directions. It may be invisible to the eyes, but we can feel its gravitational force. It passes through all matter in the Universe, including human beings, as though it weren't there at all... if it has even the tiniest hint of an ability to collide with

either normal matter or radiation, we'll be able to detect it." (Siegel, 2018)

The omnipresence of Dark Matter further supports the spiritual ideology that everything is connected. This substance may not be proof of universal consciousness, but it is proof that the Universe is more deeply impacting our lives more than we think. We can detect these particles when the dark matter collides with our normal matter. According to many spiritual ancients, energy can be disrupted or changed, so long as the user understands how. Therefore, it may leave us to wonder if this movement of particles and energy that runs through us, maybe the key to becoming consciously connected to universal energy.

Melan/o is a root word in the English language that is sourced from Greek/Latin.
The *English* language is a pigeon language that is made up of a variety of other
languages. *American English* is, in fact, a pigeon-pigeon language, as it was restructured from the original pigeon language. The structure of English is similar to that of Germanic and Latin languages, in which root words and prefixes are combined to suffixes to create new words. These root words and suffixes are binary code, that we can mix and match to formulate new descriptions. Melan/o is a root word that means *"black."* The suffix *"-in"* is used to formulate words in chemistry, which

tends to refer to a neutral substance. Thus, *"Melan-in"* is a black chemical substance that exists within us. This is not referring to dark matter, as described earlier, but it is referring to a natural black substance that is embedded within us, which may be another clue to altering our reality.

Passive Aggressive Racism, by the way, is a tactic in which racist actions are indirect but present. A common example of passive-aggressive racism can be illustrated in a case where a "willfully ignorant" white person, chooses to interrogate a black person who uses the word "melanin or melanated." Melan/o is a part of the English language; therefore, a well-versed person would know the meaning of the word Melanin without inquiring. Thus, when a black person is interrogated about this subject, it is not because the person is unaware of the meaning but rather, attempting to create confusion and doubt. The word Melanin is not newly formulated, as it was created in the year 1853.

The American English language became one of the mechanisms used to produce and maintain racism. Numerous words are used to formulate this pseudo concept that white is docile and black is extreme. Melan/o is just one of the root words used to create these concepts. Words like *melancholy* (which means "a state of dark emotions") or *melanoma* (which means a

malignant dark tumor of the skin)
or *melodrama* (which means a dark, pathetic drama); are all words that are formulated with the ideology of black vs white.

With that said, to break the spell (words are a form of spells, see the section: **Symbols**) we need to focus on the excellence of Melanin.

When we look up at a clear night sky, we can get a glimpse of the black universe that surrounds us. According to our physics, the absence of light is darkness and the absence of white is black. The glue that holds this universe together when viewed closely seems to be invisible. However, on a grander scale, the universe is more melanated than any living creature on this planet. Conceptual science may change the term to seem anything but black. However, on a more straightforward perspective, it is undeniably black. This is important to understand because we, as living beings on this planet, need to embrace our core genetic substance… we need to embrace our blackness, to understand ourselves.

The social structure that we have developed for ourselves as our social norm, contradicts the cosmic law and cosmic importance to our individual and communal identities. We are taught that we have to always choose between one or the other. In this society, one thing must be right and other things must be wrong. This concept of pure good and pure evil is our weakness and greatest barrier to finding peace for ourselves. This illusion keeps us vulnerable to remain impressionable to the ideals of influencers, stereotypes, dogma, and stigmas.

A dear friend of mine pointed out that a person can dislike the persona that a music celebrity may represent but may still be a fan of their music. A person may claim that they hate vegetables but may still enjoy a salad. The propaganda about a foreign county may promote the culture's radical views but the actual country may still be diverse with different ideals in their social structure. Everyone in Haiti does not practice Voodoo. Everyone in Jamaica doesn't hate the LBGT community. Every African American isn't poor, and thugs. And, every white person is not racist. Every voting American doesn't only embrace one political party over another, in many cases, people find political views that they can relate to within each political party.

Not having a one-dimension viewpoint, does not make us outcasts... it makes us human.

We have been conditioned to fear the darkness or blackness or "unknown"; when we should instead cradle to concepts of diversity so that we may be able to further evolve and develop. After all, if darkness was so bad, then why was the creation and life formed from it? There are so many layers to this rabbit hole called Creation.

A layer that tends to get overlooked in our creation story, is that we are all connected... literally. We all share the same Dark Matter particles that move continuously throughout the different realities of this universe... like an invisible nervous system. It is not farfetched to connect shared consciousness or thoughts or wave patterns on this river of black energy. And thus, it would be a major clue for science to undercover how we think and move similarly without being aware of our actions.

We move like how millions of particles and neutrons move in our bodies. Even the structure of our cities imitates the structures of space and molecules. Everything follows the same pattern... unconsciously.

Dark Energy, similar to Dark Matter, is invisible to the electromagnetic spectrum. It is scientifically defined as the energy that causes the universe to continuously expand. Dark Energy makes up the remaining substances of the universe, which is about 70%, making is a cosmological constant which is also calculated as being the equivalent of zero-point radiation or energy (ZPE). In Quantum Mechanics, ZPE can be found in various areas of matter (Matter Fields – Fermion Field) and energy (Force Fields – Boson Field). Zero-Point Energy (or Dark Matter), is the point of complete neutrality where positive energy (boson field) and negative energy (fermion field) cancel each other out.

The concept of neutrality (which is further analyzed later on in this book), plays a major role in spiritual science and can be found in countless religions that were established throughout human history. The notion of "As Above, So Below" could not possibly ring anymore truer. In the universe, the greatest factor is an energy that is only found at a point of neutrality. In spirituality, the only time a mere human can reach the highest level of awareness is by reaching a point of neutrality... or in other words, when a person finds their point of neutrality, or are perfectly balanced, they are connecting themselves to the Dark or Black Energy inside and passing through them.

> *"Great state I'm in, in all states I'm in, I might final form in my melanin.*
>
> *Black Power! Louder! Black Power! Louder! Black Power! Louder! Black Power!"*
>
> – from the song Final Form by Sampa the Great (Released 2019)

Spirituality

On a less mathematical ideology, the cosmology of spirituality is that the universe and all living creatures are connected. The notion of spirituality is formulated around the ideology that there is a shared consciousness between all living beings. Additionally, the fate of all humans is to obtain balance and wisdom so that they can ascend to the next level.

In the book called Yurugu, the author Marimba Ani describes how the popular scientific notions present information out-of-context, so that the concept may be presented in an Objective format. This abstracted concept can then be presented with other "non-related" ideals, to persuade how the information should be interpreted, making this concept an Objective Truth. This style of delivery of content causes the impression to appear "different" and/or "more logical" than the source. Thus, making the Objective Truth seem more real, while making the source seemingly a fable.

For example, our ancestors described spirituality as a shared consciousness, which was usually depicted as the essence of a deity being placed within all creatures. The scientific impression of this ideology can be found in various terms such as Dark Energy, Dark Matter, Zero-Point Energy, etc. Ultimately, the concept is the same story that has been told or presented differently.

Marimba Ani was able to define this process of developing Objective Truths, in a simple formula, which is:

Cosmos - Spirit [aka source of origin] = **Objective Truth**

The Spirit of this subject can be characterized as the emotional attachments or connections, associated with the subject. We can better understand this by examining synthetic substances. All synthetic substances originated from a natural resource. By isolating specific properties from a natural resource, scientists can create new materials. In some cases, the abstract substance is better than the original resource because other properties associated with the resource could be damaging. In other cases, creating an abstracted substance is not as beneficial as the natural resource because the natural resource may have other properties that would be complimentary. And in rarer cases, the abstracted substance is neither better nor worse than the original substance. For example, the Poppy plant had been used to relieve pain, help with sleep and ease digestion but the side effect is that it is a highly addictive narcotic. One of the earliest references of the opium poppy is roughly around 3,400 BC in Mesopotamia. Around that time, the plant was being cultivated in the lower regions of the area. For centuries, there was a giant push for opium, which leads to the drug wars of the 19th Century. Western Science eventually began separating the beneficial properties of the plant, so that it can be used to help people while attempting to reduce the possibility of addiction. Synthetic derivatives, such as morphine, codeine, oxycodone, and heroin, were created. In the mid-2000s it was reported that these derivatives are just as highly addictive as the original substance.

Even though the outcome of the full story may not be positive, it is still important to understand the full story. Another example of Objective Truth would be the craze in conducting a DNA ancestry-search. There has been an increase in getting your DNA examined to check out your ancestry. This is a very interesting trend that I found, similar to the Black Lives Matter statement, remains open-ended and incomplete. After all, to obtain a printout of your percentage of the possible mixture, does not tell an individual how they came to be mixed with these things. I have had a few individuals happily tell me that they have 6% African in their genetics, which is why they are taking an interest in African cultural items. I continue to let these individuals carry on, care-free, but in my mind, I contemplate about how ironic this process is... to romanticize historical torment.

I also understand that it could be argued that by becoming aware of the Objective Truth of this individual's origin, this individual could choose to become more sympathetic towards victims of racism. In an ideal reality, that would be the case, where globally we would stop colorism and racism after witnessing the results of these genetic searches. The actual reality presents a different result of the action, which is that these individuals with a very little percentage of African can feel justified to commit cultural appropriation for mere amusement, instead of for trying to make the world better.

Sometimes we are trained to only be able to recognize the abstracted content. When you have only presented the Objective Truth, you will naturally doubt the existence of any other truth. To put it differently, the foods that we buy at our local grocery store are the abstracted content from the actual source. In early learning, children are taught that pineapple is the fruit that you see at the grocery store. There was a short video that went viral in 2018, which was about showing people where their food comes from, the viewers' reactions to seeing the actual source of these foods, was very interesting. Many never even thought or even conceived that a pineapple would come from a rooted plant instead of a mysterious tree. Or that that less appealing vegetable called Asparagus, is a fern.

Imagine if we were shown the full pictures of these foods, then we would naturally understand the healing benefits of specific plants. If you understand the nature of the source, then you can identify how this creature evolved the way it did and formulated its defenses the way it did. Or in other words, you would understand the definition of these plants and minerals, and thus, you could become a true natural healer. We would also naturally understand why some synthetics shouldn't be mixed or created or used in a specific way... we would probably overcome Cancer if we allowed ourselves to understand the full picture instead of a fraction of the story. Even in our language, to put it differently, if we understood the formulation of words and memorized the roots and suffixes, then we would be able to identify the true meaning behind statements as well as be able to naturally identify --- bullshit.

Sometimes we need to revert to the source to better understand the Objective Truth. By reverting to the source, we will begin to understand other aspects of the context such as the reason behind its flavor or even the reason behind its nutrients or the reason behind its tone.

Objective Truth + Spirit = Cosmos (Truth)

The more we understand the source of the context, the more we develop the bigger picture in our minds. And the more we can understand why somethings are healthier for you than others. Additionally, the more you understand and can perceive the bigger picture, the more you will understand what spirituality is and how it plays a major part in our lives.

It is not that "objects, places and creatures" have "a distinct spiritual essence"; but rather these "objects, places and creates" share the same spiritual essence while maintaining their presence. Nothing is dead. Everything is built with living atoms and particles that are constantly active, changing and influenced by the energies it gets exposed too. By recognizing that there is a bigger picture of your story, the more you will be able to perceive and understand the notion that we are all connected to a singular spiritual essence, which is called the Collective Conscious.

In neuroscience, we learn that we are born with a natural ability to absorb information, as do other species on this planet. It is through our social and cognitive skills, that we can learn and develop, as well as to help develop others that surround us. Our ability to interact is one of the keys to our evolution. Our entire bodies, as well as the bodies of the other beings on this planet, seems to be designed to absorb information and to collect data related to their environment.

Spirit Science states that we are not only capable of absorbing information on the physical environment but also, beyond our known surroundings. The theories in spirit science seem to explain one of the "unsolved" mysteries, within Neuroscience, which is the understanding of how humans evolve on a global scale. It is an amazing phenomenon, where one group of people, in Australia can affect the behavior and cognitive skills of another group of people, in the UK. Both groups have absolutely no interaction with each other, and yet, when one group finds the solution to a complex problem, the other group suddenly can understand the same solution that they didn't understand before.

The answer is so simple... we were built to absorb information on a celestial level, as well as on a physical level. By making this assumption that our bodies can pick up all forms of data, we can understand the other scientific anomalies in this reality. An example of a popular anomaly would be the mystery behind animals being able to sense the unknown. Or why we see or sense things that don't seem normal. Or even when people seem to be able to predict things before, they happen. These strange occurrences aren't calculated in the final equation within western Science. According to the skeptical scientist, there is a so-called logical explanation for everything. But this logical explanation that many skeptics are seeking, are merely abstracted contexts that dance around the source, only to end up providing an overcomplicated answer for the same thing.

Our Ancestors believed in the connection between this reality and others. Additionally, many of them believed that time is infinite, there is no start or finish, just constant movement of frequencies. Everything that we can physically see, is made up of particles. These particles are constantly moving and causing friction which generates more energy. Our Ancestors had already understood what we are on a metaphysical level, hundreds of thousands of years ago, and had called it Atum before it became Adam, and then Atom.

On account that everything is made up of Atoms, you should take note that everything responds to frequency. Everything living being carries its personality... of sorts. It is but a very basic form of awareness that may or may not contain complexed intellect. But there are small actions that suggest, that on some level awareness in some of these beings, exists. For example, if you sing to your house plants, they respond by growing faster and stronger. If you scream at your house plants daily, or your house is filled with screams and arguments, then your house plants with receive that negative frequency will begin to wither. The evidence of this was proven in IKEA's Stop Bullying social science campaign, which lasted for 30 days (May 10, 2018, to June 9,

2018). Izzy Kalman from Psychology Today writes in the article titled IKEA Bullying Experiment Promotes Fraud and Fragility (May 17, 2018), that:

"In a school in the United Arab Emirates... one potted plant is subjected to a non-stop recording of complementary, supportive statements while an identical plant is subjected to an insulting, hostile recording. After 30 days, the complimented plant has flourished, while the bullied plant has wilted."

For skeptics, this may not be enough proof to determine if plants are affected by the tone of frequency in their environment. I recommend that everyone should try this experiment for themselves, to end their skepticism. IKEA's experiment published how words affect people. And indirectly, this experiment proved on some level, Animism exists. After all, if those plants were *"lifeless matter"* than the planets would be unaffected by the environment, other than water, soil, and light. It could also be argued that since the plants were made up of mostly water, the frequency produced by negative speech, disrupted the plant's autonomy, which caused the plant to wilt. If that is true, might I add, then that would further support the power that frequency has on the body, as humans are made up of 80% of water? Thus, negative frequencies would disrupt our body functions, just as negative frequencies affected the plants.

Spirituality is the belief system that consists of animism, ancestral worship, astrology, astronomy, quantum mechanics, neuroscience, psychoanalysis, and metaphysical science. The concept is that you not only acknowledge that most things and creatures are filled with living energy, but you also recognize the complex nature of creation and how we are all connected. In the natural world, spirituality is instinctive. Animals, plants, and objects are probably more in tune with these connections than humans. There is power in simplicity and balance. The inability to grasp the power of simplicity as well as the determination to over complicate the obvious answer is the result of ego.

There are two types of physical awareness of the self that affects us in this reality. These are the **Egotistical Self** and the **Eternal Self**.

- *Egotistical Self:* is the persona that stems from our subconscious, which is self-obsessed.
- *Eternal Self:* is a persona that lies deeper than our subconscious, on the Seat of the Spirit. This persona is also referred to as the inner child, who is free-thinking, confident, happy, at peace and aware of everything. This is the most powerful persona of self, and it is the version of

ourselves that we strive to connect with or should strive to become.

Levels of Awareness

As above, so below.

In the metaphysical ideology, there are many different variations and dimensions to "reality." These different layers are reflected in our subconscious as well as the world around us. We must understand where we are going and what we will face on our journey to finding our Eternal Selves. In this unique adventure, we will come across several distractions and possible detours that could potentially put us through spiritual turmoil. Therefore, we need to be somewhat prepared before we make our first step.

Before we get started on our first adventure, we need to first study the route we are going to use. Our consciousness is broken down into four sections:

- **Superficial Self** = *physical awareness (the person you see in the mirror and can physically touch)*
- **Subconscious** = *surface subconscious*
- **Super Subconscious** = *deeper consciousness that creates creative thoughts and concepts*
- **Eternal Self** = *divinity*

The terrain of our subconscious can be wild and difficult to navigate through. Like any experienced explorer would tell you, it is important for you to get familiar with the terrain and environment of these realms. In spirituality, you can do this by examining the characteristics of each realm. Luckily for us, our ancestors created maps with descriptions of each realm. The **Tree of Life** is a representation of the connection that humans have to this planet and the spiritual realms. It symbolizes the mapping of our inner selves as well as the mapping of the spiritual planes. It also symbolizes the mapping of our subconsciousness and the universe.

There are several different versions of the Tree of Life, throughout human history. Even though the civilizations and languages seemed to be very different, the description of these maps was extremely similar. For this chapter, we will explore two versions of the Tree of Life: *Kabbalah and Kemetic.*

On the Kabbalah version of the Tree of Life, the realms of awareness are:

- **Assiah** = *the Kingdom (physical world)*
- **Yetzirah** = *World of Formation*
- **Briah** = *Creative World*
- **Atziluth** = *Archetypal World*

Meanwhile, the Kemetic Tree of Life has the similar structure, except for the addition of the layer of Anu, that lies in between Duat and Nunu.

- **Ta** = *Physical World*
- **Pet** = *Air/Heavens*
- **Duat** = *Intellect/Wisdom*
- *Anu*

- **Nunu** = *Divinity*

When we put each concept next to each other, to compare, you will notice that each type of awareness, means the same thing. It is also important to note that in the list below, each level of awareness has been paired with an ***Elder God***:

- **Superficial** = *Assiah* = *Ta* = *EARTH*
- **Subconscious** = *Yetzirah* = *Pet* = *AIR*
- **Super Subconscious** = *Briah* = *Duat* = *WATER*
- **Eternal Self** = *Atziluth* = *Nunu* = *FIRE*

In Buddhism, the characteristics of the seven chakra points in our bodies also imitate the nature of the dedicated elder god which is also replicated with the various levels of the Tree of Life. Understanding the characteristics or the nature of the level of awareness will help you understand how to navigate through these stages. Furthermore, it will help you to understand where to look in your meditation, when things are not right or feel imbalanced, which inevitably will help you with uncovering solutions to your problems and in a sense, turning you into a self-efficient deity. In the Appendix section (and later on in this book) I explain more about the role and characteristics of the Elder Gods. But for now, here is a summary of their traits:

- **EARTH**: *Skeptical, practical, seeing is believing and grounded in physical reality.*
- **AIR**: *Indecisive, free spirited, impractical, restless and doubtful.*
- **WATER**: *Creative, emotional, empathetic, intuitive, overly sensitive and get easily emotional caught up with everything around them.*
- **FIRE**: *Inspirational, passionate, willpower, determination, impulsive and irrational.*

Levels of Consciousness (Layout)

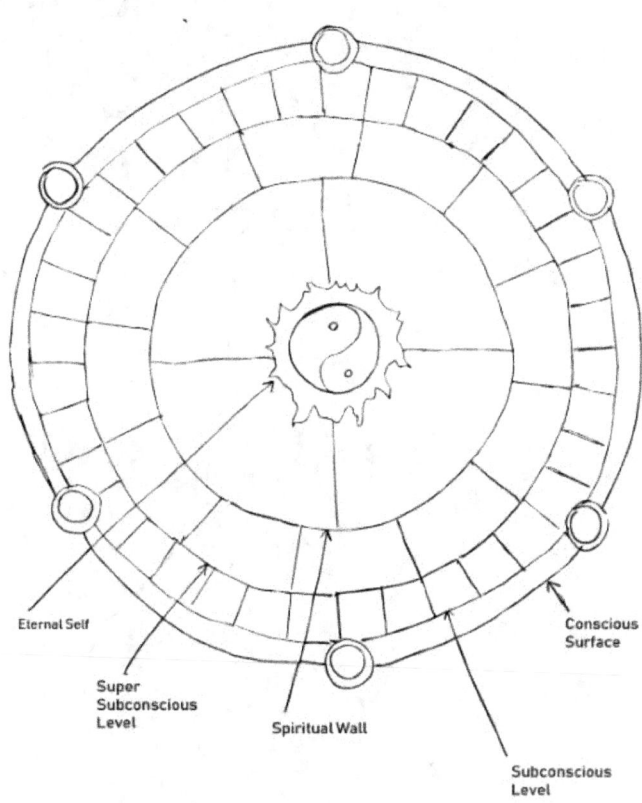

Eternal Self
Super Subconscious Level
Spiritual Wall
Conscious Surface
Subconscious Level

Tree of Life (Layout)

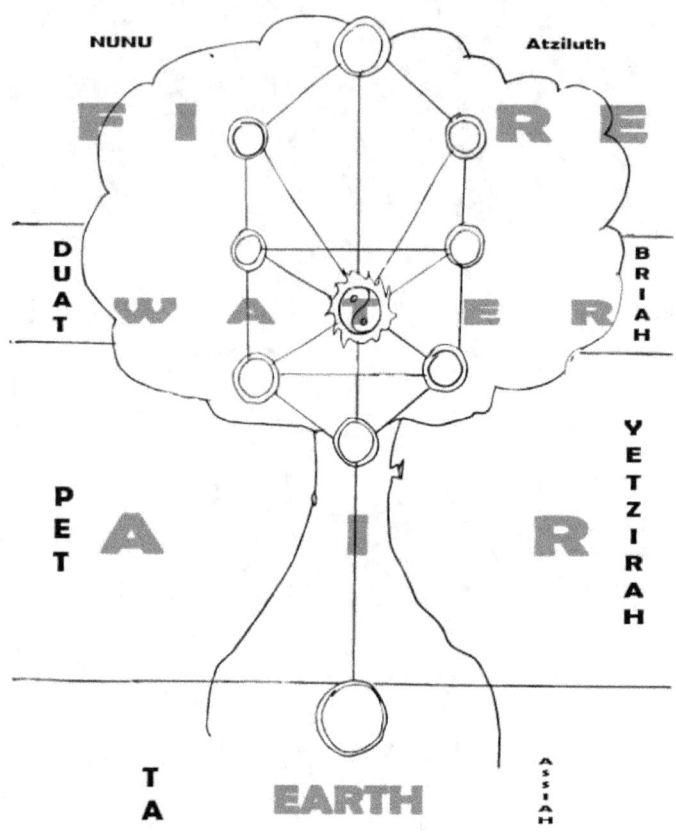

Spiritual Level (Layout)

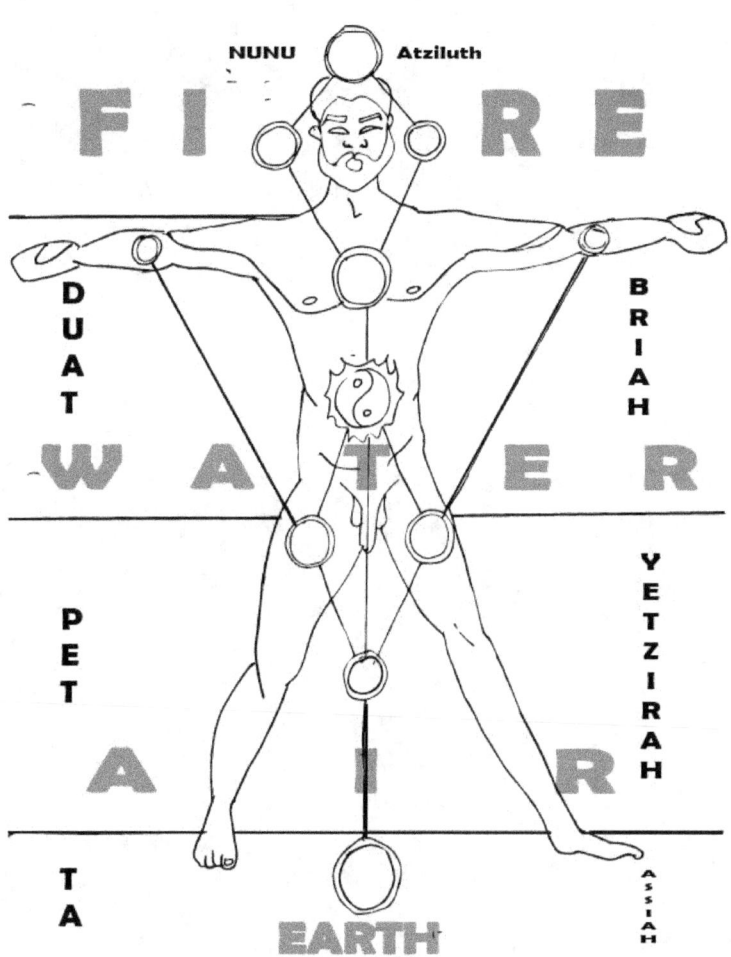

Introscoping for First Timers

When you are meditating for the first time, it can be hard, because you are bound to the physical plane, Elder God of EARTH, of awareness. It will probably be difficult for you because of your skepticism. Therefore, you must try to take a leap of faith, into the unknown of your consciousness. For the first time, it may take a long time for you to sink passed your superficial awareness and to allow yourself to relax into your subconsciousness. In the subconscious realm, Elder God of AIR, it is so easy to get distracted and lost. Images are coming and going so quickly, that you find it difficult to pay attention to one thing or another. This is the region where you formulate doubts about your decisions. While you dwell here, you will begin to forget what you were trying to look for. Other thoughts will randomly try to get your attention and you will begin to lose focus. You will have to use your willpower to stay on track and to move on to your super subconsciousness.

In the super subconscious realm, Elder God of WATER, this is where your imagination will start to create beautiful ideas and images. You will see the most amazing and beautiful things here. This is the realm where many creative ideas are formed... and lost. For the reader who is an artist, I am sure you have unconsciously visited this realm a lot. This is the place where that amazing song was formulated. Or painting. Or design. Or the story. And once you saw it you knew that this vision was so breathtaking that you needed to recreate it in the physical realm... except, once you try to bring the idea out of the super subconscious, it begins to diminish and fades away in the subconscious. So that by the time you open your eyes, you only have fragments of the vision. Returning to the super subconscious, you will also find that a lot of your emotions dwell in this place, too. For those who are battling emotional experiences, this may be the hardest place to come too. Just as you can be swept up by the breathtaking images, you can become trapped by the shadows of your memories. Be wary that you do not get swept up in the anguish of your Past. There are so many beautiful and emotional things here, that can take you off your path. Just like before, you will need to use your willpower to stay on track and move along to the Ring of Fire or the spiritual realm that surrounds your Eternal Self.

The Elder God of FIRE guards the realm of the Eternal Self. This realm is like a mirror that reveals your true character. If you are not true to yourself, then you may fear moving forward. The power of the spirit is so intense that it will turn even the bravest into cowards, frozen in their self-doubt and denials. You will need the courage to pass this gate into the realm of the Eternal Self.

The Seat of the Spirit is where you will find your Eternal Self. The image that you will see here differs from person to person. It is a very personal place and experience, that even you may not be consciously aware of. But it resides in the core of your being or the seat of your spirit. It is the point where your life force energy comes from. At this seat, your Eternal Self has a powerful place to see everything and can record all of your experiences that you have had in this life and others. It is inspirational to finally see and even more, moving to see your Eternal Self face to face.

We each will perceive our inner selves as best as we can understand it. However, Blackness is at the core of it all. Throughout history, you will find images of holy figures with black faces. These images can be seen in the Ancient Americas, Africa, Asia, Europe and so on. Our ancestors viewed the blackest of characters to be the most powerful of characters. In Russia, the medieval Christian characters are black, and there are even images of a child on the lap of the Holy Mother. This child seems to be holding a pitch-black circle that appears to be a black power coming from the seat of the spirit.

How excellent blackness must be, to inspire such powerful works to be illustrated.

In short, there are many levels of awareness. Each level of awareness imitates different characteristics of Elder Gods and emotions within ourselves. Once we pass the depths of our subconscious, we will be able to find the answers to any question and find the strength that we need to move forward. God is literally inside of you, but you need to move passed the skepticism for you to finally see It.

The Importance of Identity

There is a pattern in history where our ancestors placed a high value on maintaining their identity. When colonialists came ashore, the first thing that they did to take over land was to first destroy the identity of the native peoples. They destroyed the native identity by:

- A. **Education**: the colonialists "reeducated" the native peoples by renaming the local plants and minerals in their foreign language and forcing the natives to learn this foreign language and to stop speaking their own. Meanwhile, the native libraries were ransacked and then destroyed.
    - a. When you change the name of a Rose, you have the power to control how sweet it is.
- B. **Religion**: the means to "civilize" native peoples was through missionary work. Children were separated from their parents and raised like orphans. Parents were forced to convert to get access to stolen privileges, and so on.
    - a. When you tell a civilization that they are uncivilized because they do not follow a foreign practice, then regardless of the civilization conforms to this ridiculous

notion, they will forever be marked uncivilized by the foreign standards.

C. **War**: If the natives refused to conform to foreign religion, laws, and education; then the next step in colonialization was war. In this step, it was anarchy, where the attackers felt justify to rape and kill and kidnap.

   a. When you string an innocent man up by his thumbs, after raping him for weeks and practically starving him; you can break any strong-willed man or woman. Eventually, they will conform in hopes to end the suffering.

The continent of Africa was isolated and protected from foreign invasions for over 10,000 years. But when internal civil war broke out (the Upper Kemetics against the Lower), the defenses were weakened. Meanwhile, as the climate continued to change into a desert, resources were becoming scarce and it forced the people of Kemetic to relocate to other regions of the world.

Kemetic was left vulnerable for foreigners to take. Additionally, the native people superstitiously believed that ghosts roamed Kemetic and feared to return. As time went on, the stories told by the tribal elders became exaggerations and legends.

Once we forgot who we were, we were already defeated. It was only a matter of time before other nations either fell or began to sell each other out to foreigners.

The Americas also experienced a similar fate, in which many of the native tribes of the region were also fighting their civil wars and fighting for resources and power. When some of the native tribes of the Americas became distracted with civil disputes, they weakened the defenses of the region, just enough, to make America vulnerable for attack.

"On the Western Plains, pre-Columbian warfare—before the introduction of horses and guns—pitted tribes against one another for control of territory and its resources, as well as for captives and honor. Indian forces marched on foot to attack rival tribes who sometimes resided in palisaded villages. Before the arrival of the horse and gun, battles could last days, and casualties could number in the hundreds; thereafter, both Plains Indian culture and the character and meaning of war changed dramatically. The horse facilitated quick, long-distance raids to acquire goods. Warfare became more individualistic and less bloody: an opportunity for adolescent males to acquire prestige through demonstrations of courage. It became more honorable for a warrior to touch his enemy (to count "coup") or steal his horse than to kill him." Drake, J. (2000). *Native American Wars*.

White history will teach us that the native tribes of these regions were uncivilized and uneducated barbarians and due to this reason, they were easily persuaded to surrender their land and lives to foreigners. But that is a lie. Neither the native Americans nor native Africans were "uncivilized." All of these nations were very sophisticated, hygienic and had a self-sustaining economy. These regions also had libraries. Knowledge, in the ancient world, was more valuable than gold. It was very common that even in the most primitive civilizations, there was still some form of record-keeping and library system. Larger civilizations had giant libraries, while smaller villages had record-keepers like their Elders or Shaman.

One of the first steps in conquering people was to steal or destroy their records, Elders or Shaman. The native American and African libraries were burnt. While their priests, elders, and shaman were executed. The colonialist style of conquering other people or a previous civilization can be found in their religious texts, where their Nordic gods would slaughter their elders and steal their resources to gain power.

My point is that it is important for you to mentally throw out the lie that people of color come from a lineage of barbarians. People of color come from a long lineage of brilliance and sophisticated civilizations. Separating yourself from the lies that detach you from learning more about your family history, is extremely important.

With this said, when you are researching your history, try to understand and learn about ALL of the different factions that make up who you are. It will help you to be able to communicate with your Eternal Self. Therefore, understanding your story will give you the clues you need to ascend. There are no pure races in the modern world. We are all mixed with various cultures. Fancy DNA genetic history testing may say that I have 5% European in my genes. But from this result, I cannot assume that that mixture came from a love story of a slave with its slave master. I must assume that there is a strong possibility that my mixture comes from violent experiences. I must also be realistic with the understanding that not all stories are bad, there may also be powerful clues with identifying who my ancestors are. In short, the more you know about your earthly lineage, the more powerful of a connection you have to your ancestors.

Sacrifice for the Ancestors

For people of color, it can be very difficult to track your lineage passed your great grandparents. It is not only difficult, but it is frustrating and depressing. Most yearn to know where they come from and being denied that story due to enslavement, is a hard thing to face and live with. Our ancestors, however, are beautiful spiritual light beings. They want us to learn about them and they want to become closer to us.

In my family's old Baptist church, they used to say that:

"If you take a step towards God. Then, God will take two steps towards you."

I feel that that statement holds true to your ancestors, the Orishas and the Divine. They all eagerly want you to know them. But the only way that you can know them, is by knowing yourself. It is the ultimate sacrifice to the ancestors which is taking the time to discover who they are. When you start taking the journey to find your story, you will notice that things will start falling in place for you to find. I cannot guarantee that you will uncover their names, but you will uncover their story, and that is pretty powerful.

With that said, you should create a shrine for your ancestors. The Shrine can be located anywhere that you feel it won't be disturbed. This shrine can be decorated with things that you know your ancestors would like, such as:

- Favorite food
- Money
- Water
- Orisha Statue
- Favorite drink
- Pictures
- Incense
- Candles
- Books
- Perfume
- Plants
- Music (CD, Record, Cassette)
- Ashes, etc.

Additionally, you should also have glass or jars filled with water. This is because Water is an element that connects us to the spiritual realm. Once you have created your Shrine, you are ready to conduct your daily libations.

Prayer to the Ancestors
"I offer Light, Love and Respect; as I pour the libation in honor of my ancestors, to those whose names I know [Speak each of their names] and to those whose names I do not know. To all my ancestors I offer to liberate and uplift their souls through the Divine.

I seek mutual protection, as I offer protection to my ancestors who seek light to guide them through the darkness, I also seek protection from my ancestors as I survive through this world. I seek mutual love and comfort, as I offer compassion to the troubled souls, I also seek compassion for the days when I need comfort. May our destines be forever intertwined as we ascend. May you rest in power."

Libations

Giving liquid offerings to your ancestors and deities is a practice called a **libation**. Many modern religions use this practice, including Christianity's Easter rituals. This practice historically uses libations to connect to your ancestors and spiritual protectors. Libations are a very empowering ritual that helps us hold value to our ancestors and lineage.

With that said, there are cautions that we should be aware of, which is that we should not pray (or give our power to) foreign deities. A **foreign deity** is a deity that is not associated with your direct lineage. It is not good to offer your prayers or offerings to these unknowns, as you may not be aware of what you lose in the prayer process. While on the other hand, you **are always safe** to reach out and connect to your ancestors.

Our ancestors exist in higher dimensions and through DNA we can connect to them able to receive spiritual strength and guidance. This guidance could be that they may help you with seeing things that you didn't notice before or help you to reconnect to

Who are your Ancestors?

Those who are considered to be an Ancestor are those in your direct lineage. If they are not in your direct lineage, then they are not considered your ancestor. Also, your ancestors have to be deceased. Even if they died recently and they were close to your age. When they die, they become an ancestor. To be clear:

> Everyone isn't related to Tu Pac, so he can't be everyone's ancestor.

Pouring Libations

When pouring your libation, you can pour it on the ground r a house plant to represent you connecting the request through the Earth (the gateway to the underworld). Keep in mind that the underworld may not be physically under us. Spiritually, to pass through to realms of spirits, you have to perform a ritual signifying that you are passing through the physical plane. Remember, the Tree symbolizes the connecting stairway to the Cosmos and other dimensions. The roots of the tree or the plant, reach down into the realms of the resting souls in the underworld.

In our conscious state, we can connect to other realms because we have physical containers that hold our ties to Black Energy. The deceased no longer have a container (a process that they may later obtain a new one, eons). Without physical containers, they are still somewhat connected to this plane. However, similar to the characteristics of Air, they are not always focused. Offerings to them and keeping them strong in our memories will help them stay connected. There are three types of solutions you can use to perform a libation. These are:

What to Offer:

- **WATER-** Soothes the restless spirit, energy, healing and used to reconcile relationships
- **LIQUOR-** Ignites a resting spirit, used to ask for protection or to be used to do a strong purification
- **WINE-** Balance between the effects of water and liquor; used to gain balance between body and spirit

An ironic example of unconscious behavior is where African Americans would pour out alcohol onto the pavement to give a drink to their "fallen homies." This act of pouring out alcohol is the ancient practice of pouring libations which was transformed into African American rituals.

It is possible that for hundreds of years in slavery, many of the African Americans forgot why they pour Malt Liquor (a more affordable liquor) on the ground. Perhaps, hundreds of years ago, the

African slaves poured cheap liquor on the ground, in hopes to be protected against the slavers. This piece of forgotten history became lost on the Africans affected by the diaspora. And the Africans affected by this diaspora forgot what our ancestors are capable of doing and what they cannot do. What they can do, is strengthen us and help us solve problems as well as push us in the right direction. What they cannot do is smite our enemies...well, I don't believe they can't. There are a study researching hurricane paths in the Atlantic and how they mirror the slave ship courses. With that said, there is also the possibility that the hurricanes simply move along the Atlantic currents. Who is to say? Unfortunately, there isn't any recorded data of hurricane activity until the 20th Century.

But it is an interesting possibility that hurricanes are the unrested spirits that were countlessly called into action from people pouring malt liquor on the ground for their fallen homies. With that said, if it is the case that we unconsciously have called Spirits into Action, then it is only logical that we attempt to remedy this act done in ignorance, by putting these spirits to rest. **Maafa** is a ritual performed by Africans from all around the world. It is a ritual of offerings to our ancestors who were slaughtered or suffered from the holocaust. This is one way to ease their spirits. Another way is by erecting monuments to honor them and by giving them daily offerings.

In short, our connection to Black Energy is strengthened through our ancestors. Our ancestors are our guardians and mentors who help us with our daily struggles. The more you know about where you come from, the more spiritually stronger you potentially become. The more spiritually strong you are, the more of an influence on the world you can potentially have.

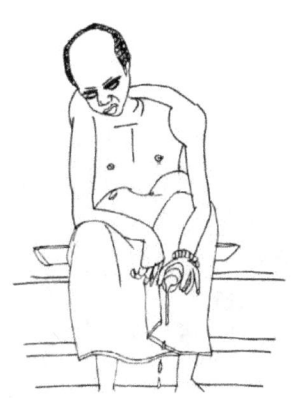

Creative Energy

Creative Energy is true freedom, that comes from the Divine and that is unconsciously weaving instructions and messages within our creations. Creative Energy is essentially feminine energy, in which creative energy is spontaneous and scenic. **Creative Energy = Black Energy = the Divine Collective Consciousness.**

'I met God and she was black.'

We can understand the ideology behind creative energy by examining the movie called *The Neverending Story*. A key element found in this film, was when Sebastian finally met the child-like Empress, in which she told him that all she needed was a new name. By him giving her this new name (*a Word used to manifest a concept into Reality*) he was bonding and accepting Creative energy (dreams) as a reality. Another thing to point out was that her destruction was happening because people's consciousness (or the collective consciousness) was detaching from **Her**. Or to put it differently, detaching from Creative Energy. Without purpose, her world began to crumble into the void (blackness) which is referred to as *The Nothing*. Once Sebastian chose to let go of his doubts and embraced this new reality; life and the creative energy blossomed and grew. As a result, Sebastian gained infinite power of countless numbers of wishes. In other words, the power to manifest things from the Divine (cosmos) to this new reality, resided within Sebastian's imagination. Sebastian held the keys or was able to connect to the Divine to make this world whatever he wished it should be (altering his reality). The more he became intertwined (connected) to the Divine the more powerful and amazing he would be. And the more amazing his new reality also became. By connecting to Creative Energy, Sebastian gained a deeper understanding of himself by exploring the vastness of this new perspective.

> Gmork, the servant of the Nothing, said: *Fantasia has no boundaries... It's the world of human fantasy. Every part, every creature in it, is a piece of the dream and hope of mankind. Therefore, it has no boundaries.*

This line reflects the concept of a collective consciousness, which is vast energy of consciousness (dreams and thoughts) without any boundaries. In reality, we face a similar conflict that Sebastian faced. This was highlighted when Sebastian's father stated that he needed to "get his head out of the clouds and to keep his feet on the ground." Or in other words, Sebastian needed to detach from the Divine and constrict his awareness to a superficial reality.

> Gmork: "*people without dreams, purpose or imagination are easy to control.*"

In other words, if we have no concept of where and what we can potentially become, then we are more susceptible to be controlled by the powers seeking dominion over free will. Meanwhile, if we understood where we came from and that this is a realm of consciousness and that we have a true purpose in our existence. Then, it would be impossible to control us because we would know better. We would know that there is a higher-order beyond this reality.

For animals like cats, this concept of living life aware of a higher ordinance, gives cats their independent reputation. One statement about the mindset of cats stated that:

> "A dog will look up at its owner and see its owner as God. Meanwhile, a cat knows better. It knows that there is a Divine being well above its owner. Therefore, cats do not worship their owners like dogs." (from the documentary called KEDI)

Real freedom lies within creativity and this creative energy is Black Energy which has no boundaries.

Divine energy desperately wants to connect to you, so that it may also exist. In the Shabaka Stone story: '*NUN created the other beings so that it may obtain an identity... a name.'* (like the child-like Empress)

But such a beautiful story of existence also has a darker side. When the Divine created a means to gain self-awareness, the product of this action also produced negative energy. Some beings rule this lower dimension. They do not want to ascend because they are content dominating this superficial reality. These beings are shadows and live under the false premise that they can alter this reality and control the Divine by controlling the collective consciousness. This act is somewhat useless because just as we are a product of the Divine, they are too. Regardless of their attempts, their actions always end up:

Unconsciously imitating the Divine or Cosmic order.

The result may not be the most accurate imitation, but it is an imitation, nonetheless. And, there are countless variations of the result. The result of how it turns out is based upon how we try to implement an action. Balance is a key factor with creation. Life thrives while living in moderation but suffers under extremities. When a creation is created out of extremities, the results are deformities.

For example, the creation of robots is a process that is pushed from Ego. The need to meet an endless capitalist demand, to obtain some form of power or self-recognition. The result is and will always be incomplete beings... hallow shells. Another example of incomplete beings that are formed from egotism, can be found in the push to create synthetic hermaphrodites. This superficial gender alteration, surprisingly, imitates the cosmic order in which 2 become 1. Or to put it differently, the act of combining masculine and feminine together to formulate a *creation*.

Naturally, through repetitive intimate interactions, two beings slowly become one. Everything about these two individuals, slowly begin to sync up with each other, from their spiritual energy to the coding of their ecosystems. On a superficial aspect, this process is imitated using synthetic hormones, plastic surgery, synthetic makeup, and other artificial replicas. Regardless of these synthetic enhancements, the coding of these individuals remains:

XX = 00

XY = 01

This is the binary language that is found throughout the universe! It is not only embedded in humans, but it is also embedded in every living being and everything; from the largest beings on this planet, to the Matter particles that formulated our existence. It is found in the frequency of the most ancient form of our languages. It is found within the farthest reaches of space. If coding was an element, 00 and 01, would be the one continuous element stretching beyond what we are capable of perceiving.

As Above, So Below.

Cosmic Order is embedded into every level of our identity. The movie called Us (2019) by Jordan Peele was about how there was an underground secret society that is tied to the actions of the world above. Regardless of how the people in the underworld resisted, they seemed to not be able to change their fate, until the day one of them escaped and changed the order of things. The key part that stood out to me about that movie was that the underworld was a very poor imitation to the world above. They were mindlessly bound, even if it didn't make sense to them because they were unable to see the bigger picture.

Similarly, we are tethered to the fate of cosmic activity or law. For the most part, we are unable to see the bigger picture. Many of us try to rebel against fate and go around claiming that nature is wrong. I do not know if nature is necessarily wrong or right. What I do know is that the unconscious actions echoing the cosmic actions are not a coincidence. With that said, Creative Energy still flows through us. We have the power to act according to our desire. It is important to point this out, too, as the fate that I am referring to, is the fate of social activity. On a more personal level, as an individual, you hold the power of your own to do whatever you are inspired to do. But just be aware that, whatever you decide to do, cosmic law will also be a part of your action, in one way or another.

One way to look at it is to review creative works. There is a lot of mystery found in art, as art is meant to be spontaneous. But in these spontaneous moments, there will often be little clues or symbols about the works that the artist is unconsciously aware of.

In my own experience, when I would showcase my art in galleries, I would often hear from a viewer about details they see in my painting that I never noticed before. Over time, I realized that I wasn't the only artist experiencing this. I found clues, that appear to be created unconsciously, in all forms of visual arts, even in architecture.

After careful contemplated, I realized that the artist sources their Imaginative ideas from Creative Energy. The result would logically be that the messages came from Creative Energy. Artists (including writers), would then be Mediums, as they are transporting these "concepts" from thoughts into realization.

A quick side story is that once when I was depressed and in Phoebus (Virginia). I painted a mural on my Aunt's studio which was adjacent to the Phoebus Library. At this time, I originally started to paint this mural as a means for me to start feeling better about myself. But as I was painting this picture, I noticed how the locals would react. You see Phoebus was an old fishing town. There were hardly any trees and the old brick buildings would span for blocks without any color. The painting that I decided to create was an endless sunset and sunrise. I worked on that mural for two weeks. During which the old oppressed neighborhood began to change. The people's body language changed while they walked. They walked around happier and greeted each other as they passed. I never completed that picture. What was interesting, to me, was that the locals became inspired to build around that painting. They built open mic shops and outdoor gatherings. People who usually drove through Phoebus to get to the floating bridges heading to Norfolk, would stop and take pictures in front of it. And when that building became too old and condemned, the locals protested that the building is protected. All of this was simply started because of a splash of color.

It was at that moment when I realized that art does not belong to the artist. Art belongs to the bystander. That experience inspired me in all aspects of my work towards black energy, including a new project called **The Phoebus**

Movement, which is a movement designed to bring creative sanctuaries to the ghettos of the world. In these sanctuaries, people will have a place to pay homage to their loved ones, meditate and to be surrounded by both natural beauty and create wonders. Every ghetto that I have unfortunately come across lacked vibrant color and natural wonders. But when I visited other regions in the same city, other cultures decorated their neighborhood with visuals to make people stop and stare.

Examining the complexity of ancient work, I truly believe that in those periods our ancestors were more connected to black energy than could ever be in our current state of mind. Their works reflected clues and information about things beyond their technical abilities. This would possibly explain how seemingly primitive tribes, like the Dogon people, we're able to predict Sirius B. In tribes like the Dogon; music, dance, and art are very important in their culture. When artists into the ***mode*** of creativity this is like going into a spiritual trance similar to a Medium or Shaman. The complexity of the ancient works demonstrates the artists put in extensive detail in their word, to the extent that modern-day artists no longer imitate. To create the beautiful walls found in Kemet or by the Mayans, the artists would have had to be in a continuous Creative Mode (or trance) to maintain focus on such intricate works. After long periods of working on their creative projects, the creation became more and more detailed as their minds expanded.

Without technical distractions and being surrounded by natural beauty, including being able to see the stars, our ancestors were closer to being in sync with Divinity, then we can do today. Now, many of us go out of our way to go camping to reconnect to nature and to see the stars, we are still not exposed to the cosmic beauty as much as our ancestors ever were.

As we mentioned earlier, every time Sebastian (Never-Ending Story) wished (or allowed himself to dream), then he became, even more, a part of Fantasia. In exchange, Fantasia became even more beautiful as it reflected his imagination.

In short, you should make it a requirement to practice or learn a new creative skill. For some, they are naturally born to be creative. For others, they have potential that they are afraid to use. Training yourself to be creative is a means to exercise your muscles in your mind(s). As you develop this creative skill, you will undoubtedly find that your awareness will heighten. The more you allow yourself to fall into the dreams (super subconscious), the more you practice bringing your dreams into this reality; the more beautiful your dreams will become, which will inspire you to go beyond your current abilities.

Spiritual Evolution Theory

Diving deeper into the concept of "Self" leads us down many different rabbit holes. One of those rabbit holes is the exploration of dimensions, time and space. For a moment, let me simply ask you:

What is the time? *Time varies upon the beholder. The importance of time is depicted by the dimension you exist in.*

The concept is that the lower the dimension the more that Time, is important. While on the other hand, the higher the dimension, the less that Time, is important. If we fail to ascend in this life, then are spirit is faced with either reliving this life again (as a different person) or the spirit would revert down to a lower dimension. Each time the spirit fails to meet a life expectancy, the lower the spirit descends into lower dimensions until the spirit returns to the void. This concept of falling into the void or ascending into the void (Divine), got me thinking of the importance and the possibilities associated with space and time and life.

If time becomes less important as you ascend, than could it be argued that your form changes to match the environment? What if these "dimensions" are referencing to perspectives?

Each perspective is a different reality for the beholder just as Time is subjective. What if we were capable of witnessing these other dimensions from our perspective, but from our perspective, these dimensions were merely a normal aspect of our reality, as (from other dimensions) our existence is a normal aspect of their reality.

With this in mind, I contemplated about what these other possible dimensions could be, if they could be easily identified as normal aspects of our reality. The result was that I was able to identify different levels of consciousness that we may not recognize as a variation of conscious perspective. Here is my theory:

> **1st Dimension** *is the realm of Microorganisms. The reason behind me identifying this as the lowest level of consciousness, was from the realization that microorganisms are technically living beings that are born, live and die. There are variations of different organisms (races), therefore, from a micro-perspective, this is an entire realm of creatures. In this realm, time would be the most important in comparison to the other levels of consciousness, because life is extremely short in this realm. Thus, is a spirit's awareness does not peak during this phase of existence, then spirit*

diminishes with the microorganism's death and lowers into nothingness (the void).

2nd Dimension *is the Animal Kingdom. There is a similar reason for this realm being identified as a realm of consciousness. There is a variety of races and species that provides different variations of that realm. Once the spirit reaches a specific spiritual peak, the spirit ascends to a higher dimension. If a spirit fails to learn and adapt to the lessons found in this state of consciousness, the spirit descends to the 1st dimension.*

3rd Dimension *is the realm of Humans. There are different variations of people (which may explain the purpose of races). In each of these variations, we hold different perspectives of the same conscious reality. Similarly, it is possible that if a spirit fails to learn and adapt from the lessons of this realm, the spirit will descend. Otherwise, the spirit will ascend to the 4th level of consciousness.*

4th Dimension *is the realm of Plants/Trees. This may seem insane, but we are the equivalent of plants. We may be able to move around more, but we* breathe in their resource of air. In this state of consciousness, time is less important as it is for humans because trees can live for hundreds of years (if not more). Additionally, they are a historical consistent when it comes to idol worship[ing], which I believe is a clue.

5th Dimension is the realm of Angels/Light Beings. Throughout history, some scriptures and doctrines describe guardians, angels, light beings, etc. One could argue that the idea of Angels is rhetoric. It may be true that these fantasy creatures are merely just that... fantasy. However, there is a repeat occurrence of these beings throughout human history. Therefore, for now, their realm would be found in the atmosphere, as these creatures are described as being bound to this planet. It would not surprise me if these beings were made of pure energy and not bound to physical forms, like humans. Either way, in this realm, Time is less of importance but not irrelevant.

6th Dimension is the Planets. This is where the skeptics throw their hands up and holler. Let me explain the madness behind this theory. Aye or Earth is one of the oldest deities. She has a core (a heart), she has blood (water) and she *probably has some form of consciousness. It would explain how life on this planet an emotional connection is, not just from disaster but through its treatment and wellbeing. From this point of view. There are different variations and Time is even less important than the 5th dimension but not irrelevant.*

7th Dimension *is Stars. Stars are a phenomenon. They are balls of energy that are mysteriously formed. They are born, live and die. The key thing to take note of is that all living beings generate Aura, this includes stars and consciousness. Their energy is so brilliant, that it gives life to other beings. When they die, they return to the void… to dust via black holes. If this ball of energy is capable of being a form of consciousness, then in this state of perspective, time is even less important but not completely irrelevant.*

8th Dimension *is Galaxies. More galaxies are being "found" or "formed" every year or so. This cosmic entity is a container to house other living organisms (similar to humans being the container for microorganisms). These galaxies orbit on their paths and continue to evolve. Time is even less important, but it is not fully irrelevant.*

9th Dimension is Blackness/Void/Divine. As we learned earlier, our galaxies are surrounded by a black halo of matter, which moves throughout every and anything. It is also the essence that dying stars return too... the black hole. This (from our perspective, at least) is the realm of consciousness where Time is irrelevant.

This spiritual evolution theory falls in line with the reincarnation and rebirth ideology found in Kemet, Greece, Buddhism, Hinduism, Taoism, and Spirituality. Particularly, the concept of humans reincarnating in animals. All levels of consciousness return to the dust particles they were formed from. It could, therefore, be assumed that these particles, which life was formed from, are black matter and infinite. This perspective brings new meaning to religious scriptures and creation stories. Especially the famous scripture:

 "By the sweat of your brow you will eat your food until you return to **the ground** since from it you were taken; for dust you are and to dust you will return" - Genesis 3:19 KJV

I cannot but point out that the term "ground" is referencing the Earth (or the Elder God, Earth)

which is the physical realm that we are all formed in. About that, on all levels of consciousness, from EARTH (Black Matter) we were made but to dust (Black Matter) we will return.

Cosmology

Historically, there have been many African cultures that worship Mother Earth. Whether directly or indirectly the characteristics of this planet seem to be reflected in many of the African belief systems. Kemetic was essentially the hub for all knowledge during its formation. Just like in modern Western Civilizations, individuals migrated to ancient Kemetic of the known world of that period. There are prehistoric cave carvings that were created by nomadic African tribes, which outline the Sphinx, like early versions of blueprints. The constructions of pyramids also seemed to have been practiced for thousands of years. There is an entire evolution or mapping of the development of great Kemetic, which demonstrates the excellence of the black people of that time. It is pseudo-science that pushed the propaganda that only aliens could have formulated these structures while dismissing the historical evidence that the people of that period practice these technics for thousands of years. As a result, many who are susceptible to this false propaganda proceeded to embed these fake ideals into their black conspiracies and ideology, to the point where the direct descendants of these Ancients doubt their capability of creating these pyramids. Many proceed to promote the lie that aliens or white people made everything. Ideally, the time will come when they will stop giving the accomplishments of their ancestors to other beings (whether fictional or real).

With that said, it is important to reference also to the Nok civilization, who is recorded as the oldest civilization in the world. Their cities were formed some 30,000 years ago within the region which is currently referred to as Benin. Their artwork is not only similar to modern west African art, but it is also is very similar to prehistoric artifacts that are found across Europe called Venus statues. Similar to other great black historical events, the Nok people were dismissed by white explorers as "vanishing" of the planet. This, of course, is a lie. The Nok people didn't vanish, they simply migrated due to civil unrest and climate changes. More than likely, they migrated along northern Africa to the middle east, through Europe and some resettled back in the Western African regions.

Early tribes were primarily nomadic. Roughly around 20,000 BCE, while the earliest Kemetic people were beginning to settle in the Eastern Northern regions of Africa, the majority of African tribes that roamed primarily in the central, south and southeastern regions of the continent of Africa. These prehistoric tribes eventually started to form a civilization hub, where each of their ideas and stories was brought together. In this new environment, stories about the deities were no longer just spoken around the campfires.

Instead, these stories were starting to be carved into the walls. From primitive carvings, the idea to create a better living environment was formed. And the carvings became more and more complex. For over tens of thousands of years, the stories of creation were written and edited and rewritten. Each new leader wanted an even amazing structure in his name. Each new artist wanted to outdo their predecessors.

This civilization was isolated from outside influences for tens of thousands of years, making it a prime environment for philosophy and development. Eventually, supremacy was formulated. This early form of supremacy was created under the concept of Amun Ra – the supreme god.

In ancient Kemetic beliefs, before the creation of Amun Ra, there was a multitude of deities that represented various characteristics of the environment. For instance, Shu is the deity of Air, while Nut's body represented the sky. Geb was the God of the Earth and Abu was the god of light. There are at least 1,500 deities that the Kemetic/Ancient Egyptian people worshipped, which is logical as Kemetic beliefs were worshipped for tens of thousands of years. Naturally, new characters would be added to the story and after a while, these numerous characters would lead to social confusion, disagreement, civil wars amongst the civilizations in both upper and lower regions of Kemetic. Eventually, the Kemetic culture became monotheistic which has been debated as being either their greatest accomplishment (outside of inventing electricity and advanced civilizations) or their biggest downfall.

The reason that the change to monotheistic could be seen as a great accomplishment because during the period of the formation of Amun-Ra (king of kings) the Kemetic people were at the peak of their civilization. In the 18th Dynasty, the Kemetic people created their most memorable wonders and structures. Regardless of this change in mindset, the belief system and the culture still reflected a lot of natural aspects from the heavy influence mother nature had on their reality. While on the other hand, the reason the change to monotheistic could be seen as the Kemetic's biggest downfall is that having a singular deity made it easy for foreign powers to manipulate the "will" of a supreme deity instead of 1,500.

Climate changes forced many tribes of Kemetic to migrate to different regions of the planet. As the native population began to dwindle, the region became vulnerable to foreign attack. The first recorded foreign (people who came from off the continent of Africa) attack on Kemetic, was in 1,800 BC (if that). Before the foreign invasion, Kemetic had at one point, expanded into the Middle East to the border of Asia. There is even proof that Kemetic expanded beyond those points, as the Pharaohs were found to have substances that were only found in the Americas and some pharaohs held a shard of Shungite, which is a mineral only found in the northern regions of Russia. As the Kemetic Empire expanded, the Kemetic Empire's ideology was heavily embedded into the ideology of those conquered regions, as we can see in the Babylonians and Sumerians' artifacts. Additionally, in history, we learn that the conquerors become the conquered. This is perhaps a result of cosmic order, in which the extremities of a nation must become balanced (which is done through diversity). It could, therefore, be assumed that when Kemetic expanded out, that they also were inviting foreigners to invade them.

During the periods of drastic climate change (where the tropical paradise began to turn into a desert), many of the African nomadic tribes moved west and settled in the western, northern and southern Africa regions, such as Togo, Benin, Nigeria, Ghana, Senegal, and more. They took with them, the Kemetic ideology which led to a rebirth of the Kemetic old deities, who were renamed and were worshipped by the newer reformed civilizations. The stories of the deities continued to develop, but the core structure of the story remained very similar.

Ancient African Cosmologies

The Ghanaian **Akan** (primarily the tribe called the **Coromantee**) believe in animism and a Supreme being named **Nyame** (divine creator). **Nyame** (who is married to **Asase Yaa** – Mother Earth) creates the **Absosom** (spiritual beings) to go forth and assist the humans on the earth. The Akan also pour libations to their **Nsamanfo** (ancestors). Humans are made with the following components: *mogya* (blood), *Sunsum* (spirit) and *kra* (soul). The Akan believe that humans each have to fulfill individual *nkrabea* (destiny). The *kra*, which was created by Nyame, will return to the land of its ancestors.

The Nigerian **Yoruba** believe in animism and a Supreme being named **Olodumare/Olorun** (divine creator). **Olodumare** creates the Orishas (spiritual beings) to assist humans on the earth. Every person has a destiny, which is to gain spiritual wisdom to ascend or return to Olodumare. The **Ori Inu** (spiritual consciousness) needs to grow and develop so that it may be reunited with the **Iponri** (Spiritual Self).

The West African **Vodun** (primarily associated in Togo) believes in animism and a Supreme being named **Mawu/Mahu** (female divinity – divine creator). Mawu/Mahu creates the **Vodun** (spiritual beings) to assist the humans on the earth. The followers of this belief system worship their ancestors. All creation is considered divine and has divine power. The human body is made up of the following components: **Se** (pronounced Say – which is the soul) and **Ye** (pronounced Yay – which is the power of the Divine). Ye returns to the land of the ancestors, after death.

The **Kemetic** people believed in animism and that a Supreme being named **Atum Ra** (divine creator), created the universe. Since the world was created with his powers and his flesh, every creation upon the earth was considered divine and was powerful. The power of Atum Ra was reformed into a human soul when humans were created. Atum Ra created the **Neter** (spiritual beings) to assist humans on the earth. The human body is made up of the following nine components: the **Ren**, the *Ba (personality)*, the *Ka (double)*, the **Shuyet** (shadow), and the **Jb** (the heart), the **Akh** (magical one), the **Sahu** **(spiritual body),** the **Khat** (physical form) and the **Sechem** (form).

Fig 1: **Ancient African Religious Similarities**

- **Divine Creator** → Atum Ra → **Nyame, Olodumare** and **Mawu**
- **Mankind was formed by the essence of the supreme being** → there is a link connected between humans and the supreme being, whether the supreme being created humans directly or indirectly.
- **Angels/Spirits created to assist humans** → Neter → **Absosom, Orishas** and the **Vodun**
- **Animism** → Creation is Divine → all creations on the planet has the essence of the supreme being and/or a spirit of their own.
- **Lifeforce Energy** → Power of the Supreme → **Energy, Aye, Obi, Asé.**
- **Ancestors** → Ancestor Worship → **Ancestors, Sunsum, Nsamanfo.**

The structure of these different belief systems is almost identical! In all variations of this concept, there is a supreme being who is beyond the direct reach of humans. This deity creates lesser deities to assist humans as they transcend to fulfill their destiny. Additionally, in many of these ancient stories, humans were formed from clay and dust (which we learned represents the Elder God EARTH also can be referred to as Matter). This concept of creation has so many similarities with not only each other but also can be found in modern religions.

On the contrary, ancient cosmology found in ancient Europe, have a slightly different concept of creation. In the ancient world, polytheistic ideology was very common. However, in many of the ancient European ideologies, there was no direct continuous connection between the chief deity and humans. The ideology of this region, during this period, was brutal, to where even the deities were constantly at war.

Ancient Greece & European Cosmologies:

Gaia (Earth) as well as three other feminine deities (*Eros, Erebus,* and *Tartarus*), emerge out of chaos and nothingness. Gaia suddenly gives birth to **Uranus** (the Sky), who then conducts intercourse with Gaia, to conceive the first **Titans** (six males and six females). At some point, Gaia and Uranus decided that no other titan should be created, though two other titans were created shortly after this decree. Gaia convinced one of her offspring, **Cronus**, to castrate his father/brother, Uranus. Cronus feared that his offspring would do what he did to his own father/brother, and so he ate every child he ever conceived, except **Zeus** (who managed to escape). Zeus, when grown, drugs his father to cause him to throw up and release the other offspring that was gobbled up. Zeus and his siblings (who were rescued from Cronos' stomach) battled against Cronos and the other Titans. When Zeus and his siblings won, they threw Cronos and the other Titans into a prison that was situated on Tartarus. Like his father, Zeus feared the same fate from his offspring and also ate his first wife while she was pregnant with his first child. The baby was born from the stomach of its mother, from within the stomach of Zeus. **Athena**, the baby, climbed out of the mouth of Zeus. When Athena escaped from the belly of Zeus, she was fully matured and dressed for war.

Similar to African cosmology, numerous deities represented different aspects of life, such as love and compassion and courage. Zeus created a series of deities, the Greek Pantheon (lesser gods). Unlike the African deities, these Gods had no real connection to humans. They were not created by Zeus to be the angels and spiritual guardians of humans. They did not aid humans to ascend. Their primary purpose was to be worshipped and out of occasional interest, these deities showed humans some form of mercy. In a way, humans were mere pests that only really existed to help keep the gods alive.

Both African and Greek cosmology share a similar story about mankind being created by a deity and formed from clay. Prometheus was the creator of humans, in Ancient Greek ideology. Prometheus formed humans from clay and stole the eternal fire which he gave to humans. This act gave humans the ability to progress and develop civilizations. Zeus was furious and sentenced Prometheus to eternal torment, in which the immortal god would have to endure having his liver eaten away at by birds, day in and day out. While being tormented for eternity, the children of man progressed, and the Greek deities took it upon themselves to seduce or rape the daughters of men. As harsh as the belief system of the ancient Greeks appeared to be, there was still some form of connection to earth (Gaia or Ge). Ge was represented in many of the Greek temples and played some form of role in the worshipping of the Gods.

In old Norse, the Germanic tribes of the 14th Century also followed a polytheistic system. In the beginning, in Nordic cosmology, there was nothing, **Ginnungagap** (the abyss). Out of this abyss appeared the realm of ice, **Niflheim** (a name that is ironically very similar to the *Christian Nephilim*). And the realm of fire, **Muspelheim**. The Abyss lay between these two worlds. Over time, the frost from the world of Niflheim and the fire from the world of Muspelheim merged in the abyss. From this fusion of elements, **Ymir** was formed. Ymir, translated to mean *Screamer*, had a very destructive nature. This asexual being gave birth to giants while it slept. Meanwhile, the frost from Niflheim and the fire from Muspelheim continued to fuse. As the ice melted another god emerged which was named **Audhumla** (a cow). The Cow Goddess nourished Ymir. As she stood and nourished the Screamer, she would lick away at the ice from Niflheim. In time, her licks revealed another god named Buri. And Buri's son was Bor (son of a god), who married Bestla (daughter of a giant), who bore Odin, Vili, and Ve (demigod deities). Odin and his siblings slaughtered Ymir. And from the corpse of Ymir, Odin and his siblings constructed a new world: Ymir's blood, was the oceans; Ymir's dead skin and his muscles, was the soil; the sky was created from Ymir's skull (which was held up by four dwarves), and the clouds were formed from Ymir's brains. At some point, Odin commits the ultimate sacrifice by hanging himself from a tree (which is the Nordic Tree of Life called **Yggdrasil**), cutting out one of his eyes and stabbing himself with a spear (an act that which hauntingly echoes the *Spear of Life* that stabbed Jesus at the cross where he was sacrificed), so that he may obtain the power of the **Runes** (knowledge or the Word) – like the Alphabet, Runes symbols are letters, see the chapter: *The Alphabet and Symbols*. It is also very ironic that Odin hung himself with rope (which can appear very similar to a snake) on the tree of life, where he stole forbidden knowledge. Curious that the tree in the garden of Eden, was also called the tree of life, where knowledge was stolen by some form of trickster. Odin chooses to later share the knowledge which he stole, with the humans that he helped bring to life.

Meanwhile, the original humans were two pieces of driftwood, that Odin and his siblings happened to come across. Odin gave the driftwood the

Breath of Life. While his siblings fashioned humans with intellect and motoring skills and clothes. After creating man and woman, Odin and his siblings placed them on Midgard and fenced the humans in, to protect the small beings from giants. These humans, **Ask** and **Embla**, began to populate Midgard with their offspring… until Chaos arose from **Ragnarok**, in which it destroyed almost everything. During Ragnarok, two humans (**Lif** and **Lífþrasir**) are instructed to hide in **Hoddmimis** holt (a wood) until the destruction resides. After Ragnarok, the two repopulate the earth.

Humans did not necessarily play a huge role in the realm of their deities. The **Elves** and **Dwarves** (spiritual deities) that roamed Midgard, were not necessarily present to assist humans. The destiny that humans faced in the afterlife varied based upon the style of death. The concept of a soul did not get adopted into Norse ideology until after Christian conversion and influences. If you died like a Viking war hero, you would be welcomed to Odin's hall called Valhalla. While Freya would occasionally welcome a few favored dead into her hall called Folkvang. If you died at sea, which is common for a Viking, you would be taken to the watery hearth of the giant named Ran. If you were neither a hero nor favored by Freya nor died at sea, then you would be taken to Hel (literally pronounced Hell), to the realm of the goddess named Hel. As you probably have guessed, medieval Christian missionaries adopted this term to describe the realm of Satan to the Germanic tribes. After a while, the concept of hell was incorporated in modern Christianity. The Nordic Hel was simply a continuation of life but in a different realm. It wasn't a land of torment, as the Christian hell. But it also wasn't desirable or as exciting as Valhalla. By studying the premise of these ideologies, we can detect the similarity between Greek and Nordic cosmology.

Fig. 2 <u>Ancient European Religious Similarities</u>

- **Creation has no purpose to its existence**
- **Deities formed from Nothing and/or Chaos** → Chaos, Nothingness and Ginnungagap.
- **Asexual Deities** → Gaia and Ymir
- **Age of Monsters** → Titans and Giants
- **Offspring murder or attack their predecessor and/or offspring** → Cronos castrates Uranus; Cronos eats his children; Zeus imprisons Cronos; Zeus eats his child; Odin slaughters Ymir.
- **Earthly Spirits have very little connection with mankind** → elves, dwarves, nymphs, etc.
- **Not Animism** → the ideology in these belief systems was that the actual objects (trees and plants and animals) didn't necessarily have spirits of their own or that they were spiritually connected to humans. But rather, that these objects

were managed by spiritual deities. Thus, it is not truly animism but more polytheistic.

- **Supreme being has little to no direct influence on human development and/or creation**
- **Supreme being and other deities, sexually attack and abuse lesser beings (including humans)**
- **Very little importance on the human soul and/or afterlife** → when humans die, the majority were sent to realm that holds very little description or importance to human achievement. The goal seemed to be to become favored by the Gods in order to be invited to the Godly halls. The biggest dread in this ideology, seemed to be sent to the halls of a "commoner." After death, the greatest accomplishment was materialistic glory.

Ancient Mesopotamia Cosmologies:

The region of **Mesopotamia** is modern Southern Asia and the modern Middle East. This region was a major connection point for nomadic tribes who migrated from out of Africa; as well as for the tribes migrating to the south of Europe; as well as the tribes migrating to the west from Asia. The first mingling of various human species began in this region, some 40,000 years ago. This mixing of prehistoric genetics played a major role in the creation and evolution of modern humans. Thus,

Mesopotamia became a very diverse area that was rich with genetic perspectives. As well as home to several powerful ancient powers: *Sumer, Babylon, Assyria,* and *the Akkadian Empire.*

The doctrines of Judaism, Islam, and Christianity are sourced from various religious accounts that were primarily from the Mesopotamian religions. Correlations between the Mesopotamian religions and modern Abrahamic religions can be primarily found in the **Epic of Atra-Hasis** and the **Epic of Gilgamesh**. The Sumerian concept of creation was that the entire universe was **An** (Heaven) and **Ki** (Earth), which was more commonly referred to as An-Ki. The flat earth resided below a tin domed sky. Inside this dome was Lil (Air) and Abzu (Sea). Before the creation of An-Ki, the Babylonian Epic called Enûma Eliš tells the story of the creation of the universe.

In the beginning, there was nothing, but two deities named Apsu (masculine) and Tiamat (feminine). Through the sexual union of these two deities, four other deities were created: Lahamu (feminine) and Lahmu (masculine); then Kishar (feminine) and Anshar (masculine). The linage of superior deities came from Anshar and Kishar (sister/wife). Anshar's son was Anu. And Anu's son was Ea (also known as Nudimmud). The younger generation of deities begins to irritate Tiamat (one of the original gods). Apsu tried to calm his wife, but Tiamat remains restless. A deity named Mummu became the advisor for Apsu and Tiamat, who advises Apsu to kill the younger generation of deities as well as everything that they made. The younger deities heard of the elder gods plans to destroy them and devised a plan. Ea puts a spell on his great grandfather, steals his grandfather's crown and crowns himself, then slays great grandfather. After Apsu's death, Ea imprisons Mummu. From the corpse of Apsu, Ea creates his realm. In this realm, he (and his wife, Damkina) conceive Marduk (the Sun) in the heart of Apsu. Meanwhile, the other deities run to Tiamat (Ea's great grandmother), and demand justice for Apsu. Tiamat responds by creating an army of monsters as well as fashions up her new consort, Kingu (We-ilu). Ea hears of the plans of his great grandmother, Tiamat, and goes to his grandfather, Anshar, for help. Anshar goes to his son, Anu, and asks him to speak to Tiamat. Anu attempts to speak to Tiamat but is too fearful of her wrath, so he turns around and goes back. Ea then goes to his great-grandson, Marduk, and asks him to be his champion against Tiamat. Marduk agrees, so long as he is made the supreme god and overlord, a rank that goes above everyone, including his great grandfather, Anshar. This demand makes many of the other gods uncomfortable, but in a moment of intoxicated bliss, many of the previously opposing deities, decide to support Marduk's cause. Marduk eventually defeats his great-great-grandmother, Tiamat, and becomes the overlord of all the deities.

Marduk temporarily "rescues" Kingu from the clutches of Tiamat after he slays Tiamat. From the corpse of Tiamat, Marduk forms the sky and the star constellations; as well as the time (day and night) and the moon. Determined to create humans, Marduk slaughters the previously rescued Kingu and uses his blood to make mankind. The humans were molded out of clay and blood mixture. The blood was harvested from a god reside inside of them. Marduk then divides the deities into two sets: Above and Below. Three hundred deities were sent to the heavens, while six hundred deities were sent to earth.

The afterlife in Mesopotamia was considered to be less vibrant than the realm of the living. The netherworld is ruled by the seven Anunnaki (also known as the Upper gods in the Epic of Atra-Hasis). The Igigi (also known as the Lower gods in the Epic of the Atra-Hasis) were the servants of the Upper Gods. They were forced, by the Upper Gods, to dig out the waterways. A rebellion occurred at some point, and the Upper Gods were forced to rule the underworld and judge the fate of the eṭemmu (human spirit or ghost), while the Lower Gods ruled the land of the living. The expectations of the underworld varied from region to region (*for example* Sumerian underworld was not the same as the Babylonian). Humans' destiny was to be mortal. From dust they came to dust they were destined to return. The primary purpose of humans would seem to be that they lived and then they died.

The ideology behind this Mesopotamian concept clearly outlines many similarities that can be found in the other ancient cosmologies that we previously analyzed. Within this creation story, you can see a clear pattern of influences from various cultural beliefs. In a way, the cosmology is the testament of the diverse environment and nature of Mesopotamia, which could stand to be further evidence that various nomadic tribes mingled in this region. The Mesopotamian depiction of creation echoes the brutality of the Nordic stories, as well as the divine connection of the African stories. This would historically make sense, as Mesopotamian has always been (well since 40,000 ago) a diverse region of various human races, tribes, and species. It, therefore, makes sense, that their concept of cosmology would reflect both the harshness of the Nordics and the Spiritual Connection of the Kemetics.

Additionally, the use of a divine sacrificial lamb for the fate of mankind also is echoed in Abrahamic scriptures. Another key thing to note is that the Spiritual Beings who dwell on the earth, share the same purpose as the African concept of protective spiritual beings. These deities are more involved in the progress and livelihood of mankind, verses being a distant neighbor, like the Nordic Elves and Grecian Nymphs. The superior deity plays a major role in the development of the world of men and styles his thrown directly above them in the heart of Apsu.

Zoroastrianism is a cosmology of the dualistic concepts: good and evil. This belief system is monotheistic (supreme being – *Ahura Mazda*) and believes in the ideology of a final global cataclysm. This religion was founded and made popular by the Persian Empire, before the Muslim conquest of Persia and considered to be one of the oldest active religions. Many core components of Abrahamic religions can be found in the ancient teachings of Zoroastrianism.

The philosophy of Zoroastrianism is that Ahura Mazda is the sole creator of the Universe (though, similar to Kemetic, the supreme being was presented within a triad). The name Ahura Mazda means Lord/Mighty of Wisdom (Ahura means Mighty or Lord, while Mazda means wisdom).

Throughout the progression and development of Empires in Mesopotamia, the stories of the deities would change to reflect this historical event. Such as Marduk easily receiving ultimate power from his predecessors by taking on the name, *Asarluhi* (which is one of the fifty names of Marduk), this divine change reflected the political change of power in Babylon. This influence of political and economic affairs in religious doctrines was very common throughout the ancient world. The religious literature of the ancient world was the equivalent of modern media. Many of these stories were added to influence the masses to support political, social and environmental agendas. Due to this, any scholar studying cosmology must also comparatively study the history of that period.

Ultimately, *there is only one story.* The *exact details* of this singular creation tale no longer exist or can be found, as it has been altered and changed so many times, over the years. All we have left are fragments of countless variations of a similar tale. Each variation of the story is presented in a setting that would relate to the audience of that period. For the Vikings, the story was filled with blood, war, rape, and glory. For the Kemetic people, the story was filled with balance, power and hopeful possibilities. For Judaism, the story was filled with judgment, ritual, and guilt. For the Islamic, the story was filled with reward, servitude and piteous. For the Christians, the story was filled with guilt, revenge, self-righteousness, and incentive.

The setting, or surface, of the story, is aimed at a specific target group of listeners. But underneath the surface of these stories, lie the clues that will help us understand the bigger picture. Once you strip away the political jargon, you will see something immensely deeper, such as cosmic activity.

Energy flows through everything. On scientific bases, this can be recorded as frequency and sound waves. In Science, sound waves (or waves of frequency) were responsible for creating life. Western science (which is also a form of belief) teaches that sound waves were formulated from the random explosion (aka Ragnarok), which from its incredible force of sporadic energy, which coincidently formed complex life.

From this immense force of energy, stars and planets were spaced out far enough, to where they began to formulate individual orbiting and eventually life was sporadically formed from the bacteria (microorganisms) in water (an act that nature never repeated). From these random microorganisms, other organisms decided to also spontaneously fuse until complex life forms such as plants, animals, minerals, and humans, independently formulated themselves from spontaneous and very selective evolutional jumps. This spontaneous Big Bang act had absolutely no conscious concept or direction or intellect. And yet by complete accident (according to science) a lifeless substance coincidentally created complex intellectual beings that were capable of evolving into even more complex intellectual beings.

No matter how much I look at it, the concept makes absolutely no sense. Or rather, the western science cosmology is so objectified and abstracted from the full story, that it makes no sense. Their theory can be easily disproven by asking the question:

Can dead substance consciously produce intellectual life?

There are some individuals out there who claim that dead substances without a motive can produce complex intellectual life. If you are one of those individuals who believe that this is possible, please explain this riddle: if a "dead substance" can unconsciously create or produce intellectual life, then was this "dead substance" truly dead to begin with? We are products that were produced from life. All living creatures on this planet were born from living predecessors. Every being we are aware of came into existence through another life. There are no examples where life is formed from dead matter. Regardless if you believe lifeless energy created life or if you believe that a supreme being formulated life, one thing is absolutely true... energy and frequency are both found in everything on this planet. Both of these readings can be mathematically calculated and/or harvested. This creation story has been and will continue to be told beyond what we can perceive in this space and time.

Hypocrisy of a Warped Sense of Reality

By understanding Life Energy, Animism, Frequencies and their connections to your Eternal Self, you will be able to develop a new depth to your awareness. It is hard to discover yourself if you are unaware of your connections to this planet, universe, spirit, animals and so on. With that said, the planet is extremely important to us. We need this planet to develop in, to obtain substance and to be grounded. With that said, be wary of the Synthetic Disenfranchised versions attempting to "save the planet" in a perverse manner

One cause would be while preaching about saving the planet, they are destroying natural areas with human waste and trash, rather than demonstrate respect for the planet, by cleaning up after themselves. The hippies, for example, are a sub-cult group that was derived from European principles. The concept of being a "hippie" was to return back to living free from rules and to reconnect with the spiritual self. One of the mottos pushed by the original hippies was one-love and the love and appreciation of the planet. However, though this was the verbal motto, the actions suggested other things. One example of the statement not quite matching the action would be the Festival at Woodstock (August 15-18, 1969). This festival attracted over 400,000 people, many of whom considered themselves to be a hippie. It was originally promoted as *An Aquarian Exposition: 3 Days of Peace & Music*. The festival was hosted at Max Yasgur's 600-acre farm near White Lake, New York, which was 43-miles southwest of Woodstock. Since there was no police present, drugs were heavily circulated throughout the event. So much so, that the employees working the concession stands, would swap food for dope. Throughout the festival, LSD was promoted, so much so that medical tents were set up for individuals suffering from overdose and other drug-related symptoms. Therefore, the more accurate description of this event should have been An Aquarian Exposition: 3 Days of Drugs & Music. The true nature of the event (which was based on drugs) led the attendees of Woodstock to participate in public nudity, public sexual acts, lack of personal hygiene, lack of use of proper irrigation systems or proper waste management. Leaving Yasgur's farm trashed, by the time the event was finished. As much as we would like to focus on the memorable music, we cannot ignore the reality that Woodstock was a public display of moral decline and indecency. Festivals are on the rise of popularity, again, and have caused a similar effect like Woodstock, in which tons of plastic are dumped in a single event. This hypocritical behavior can also be found on hiking trails and mountain tops, where supposed Earth Loving hikers, leave the area trashed with plastic tents and fill the rivers up with plastic bottles.

Another example of perverse behavior would be the practice of Ecosexuals. Modern Nature Worshipping cults that are derived off of European principles, practice warped perverse rituals of vulgar behavior that is a combination of both modern hippie and medieval Greek sexual practices in which these cults use nature as sexual objects (treating the plants as a lifeless matter that can be sexually exploited). This cult believes that in order to save the planet, the members of this belief must have sex with the planet. In other words, there is a group of people who masturbate and dry hump various items in nature (trees, rocks, soil, grass, etc.) Neil McArthur from Vice, states that "the term "ecosexuality" has existed since the early 2000s when it started appearing as a self-description on online dating profiles. It wasn't until 2008 that it began its evolution toward a fully-fledged social movement..." (*Ecosexuals Believe Having Sex with the Earth Could Save It*; November 1, 2016)

One could argue that everyone is free to sexually express themselves, as they see fit. I couldn't agree more. I will add that it is the message that I find to be misleading. I cannot comprehend how an individual sexually stimulating themselves against a billion-year-old planet, could "save" the planet.

There are practices aimed to "save" the planet, seem more focused on individual egotism and less on the collective consciousness. Our ancestors, on the other hand, treats or considers the planet to be sacred. In many ancient practices, offerings are given to the deities to bless the planet. Additionally, trees are commonly considered holy, as well as animals. There are more and more reports of small tribes on the continent of Africa who has devoted their lives (becoming-animal bodyguards and guardians) to protecting the endangered species. Such as the group most popularly called the Pangolin Men. Crystal Chang from Barcroft TV writes:

> "Committed minders from the The Tikki Hywood Trust in Zimbabwe work hard to protect the species with a one-on-one care programme. The charity workers are assigned with one pangolin each, where they spend 24 hours a day rehabilitating and walking the majestic mammals so that they can forage naturally"

The point of all of this is to highlight the differences in actions for "saving" the planet. On one spectrum, the actions are more self-centered and do not seem to match the promoted intent. While on the other spectrum, the action of self-sacrifice of personal time and a change to individual lifestyle, seems sincerer. Identifying these actions are important to take note of because it can help us understand the difference between constructive and destructive behavior.

Constructive behavior produces while destructive destroys. Every action has a consequence and we have been blessed with the intellect to critically decipher which action will produce the best outcome. More than likely, the destructive action will produce negative results. While constructive action will produce positive results.

Our ancestors knew about this pattern and demonstrated their understanding of "action and consequence" by devoting much of their beliefs and lifestyles to try to obtain a balance. You see, there is a purpose in destructive action as much as there is a purpose for constructive action. Our planet demonstrates the need for both energies on a day to day bases.

For example: when a hurricane comes through and leaves a path of destruction, it seems scary. But once the destruction has cleared, the air and ocean are filtered and clearer, giving room to new life. Another example would be when a volcano erupts, like Mount St Helens. It seemed like she destroyed everything. But shortly after the chaos, new life formed. And now, Mount St Helens is more beautiful than she was before. Therefore, the need for destruction is sometimes necessary in order for there to be change. We have to decide, if our choice of action, will be beneficial in the long run.

Thus, we can ask the question: *does having one-side sexual stimulation actually benefit or help. Improve the planet's condition, in the long run?* More than likely, it does not. It does, however, cause a higher risk of infections and cultivating other germs to develop in the groin area. *Does the act of using chemical drugs help to improve the condition of our planet in the long run?* More than likely, it does not. There are some horrible side effects of using drugs, particularly chemical drugs. The consequences of using these types of substances is a negative consequence, as it is destructive, in which it causes the user to rot inside out. In both examples, we can see that the most probable outcome of conducting these practices is not beneficial... thus it is a destructive extremity.

This destructive extremity outweighs the positive counterpart. In cases of extreme actions, there will be an extreme consequence, as the weight of the action is attempting to become balanced. For example: if you sip liquid heroine called Lean, that is an extremity (especially if you are doing that on a regular). The inevitable consequence is Death, as this substance quickly does a lot of damage to the heart and the brain. If your intention is to die a painful death before you reach the age of 30, then this extreme consequence may, in fact, be beneficial to you. But if you have loved ones (like children), who rely on you. Then this is a horrible experience for the practice, if not for yourself, then for them!

> *"Drugs are bad... Don't Do Drugs!"* D.A.R.E. program

The Planet is Precious

We are the spiritual replicas of Trees. Just as a tree would die if malnourished, so will we. Just as much as we rely on trees to breathe, the trees rely on us. Everything on this planet exists for a reason, even self-destructive humans.

Trees were considered very sacred in Afrika, from Kemet to the Khoi Khoi people in South Afrika. Trees represented the portal of cosmic connection or the gateway into other realities and dimensions. Due to human migration whether caused by colonization or human migrations before colonization, the Afrikans found around the world, hold the same ideology. This includes all of the Afro Americans (from Canada down to Chile) and the Afrikan tribes throughout the Pacific Islands, Australia and New Zealand. Belief systems, like Afro-Brazilian Candombe (the western version of Orisha or Yoruba beliefs) and North American Voodoo practices.

In all of these different cultures, trees and nature hold a high value and respect that is not demonstrated in a western ideology but is interwoven in various religions of those cultures.

In West Africa, the **Oubangui** people plant a tree for a child when the child is born. As the tree grows, it symbolizes the growth of the child. If tree growth declines, people fear for the health of the child and a healer is called upon. When the child is sick it is brought to the tree for treatment. When the tree begins to fruit, the time will have come for the child to marry. Throughout a person's life, gifts are occasionally left for the tree. When someone dies their spirit goes to reside in their personal "birthright" tree.

Throughout west Afrika, forests and specific trees are protected and valued as historic symbols. Elder meetings and gatherings are held under the oldest and most sacred tree, where they talk about politics and the welfare of their community. This practice of paying homage to the trees is a practice that dates back before human memory. Sacred relics and rituals are practiced under the trees. There is a folklore of a Yoruba god, who had done something mischievous. And in order to repent for his behaviors, he meditated at the top of a Baobab tree. When he finally gained enlightenment, his garments turned white and shun like the sun.

From our ancestral perspective, the world is both living and sacred. This concept is embedded in our art and stories and belief systems. There is, accordingly, any activity conducted with ancestral approval.

The concept of the Tree is a spiritual gateway that can be found in the Native tribes found throughout the Americas. They believed that the tops of the trees, was the realm of gods and that trees were the gateway to both the upper world and underworld. In some Native American tribes, the trees were decorated with offerings and trinkets to honor the spirits of the world. The Banyan tree is worshiped in many Asiatic beliefs. Many deities throughout the world's civilizations, would go to trees to rejuvenate their power, gain strength and gain enlightenment of the universe, from sitting at the base of trees. Buddha was said to have gained enlightenment from meditating underneath the Bodhi Tree.

The Tree is a constant throughout our history! It is found also in the Yoruba beliefs, where the deities' spiritual powers were sourced from trees. The deity Shango, for example, committed suicide after he lost a battle for the village called Oyo. Shango hung himself from a tree. In doing this, he was spiritually reconnected to **Olodumare.**

For us, Animism is more than the worshiping of abstract and lifeless objects, as the European Model would try to convince us to believe. Animism is the human connection between the cosmos and individual reality, we are connected to everything. Accepting this ideology is the highest level of intellect. We have been programmed through oppression to treat the planet as an abstract and lifeless thing. But without being oppressed, the world considered the planet to be a deity and that they were a mere part of a giant infrastructure. Therefore, the European Concept of nature being an Object is not as sophisticated as we were made to believe. This abstract way of thinking is primitive as the concept is rudimentary without either historical or spiritual reference.

Try It! English grammar teaches us to write nouns in title format and "things" (aka objects) in lower-case sentence format. When you are writing about Mother Earth or writing about Trees, try to write items that you spiritually connect with in title case, to retrain your mind to accept these items as Spiritual Beings.

Tree of Life

Discovering the importance of this planet brings us back to the concept of the Tree of Life and the role it plays in our ancestral legacies. Earlier in the chapter **Level of Awareness**, we learned about how the different levels of awareness are mapped out on the ancient Tree of Life charts. These levels that represented both realms and Elder Gods are:

- **Superficial** = *Assiah* = *Ta* = *EARTH*
- **Subconscious** = *Yetzirah* = *Pet* = *AIR*
- **Super Subconscious** = *Briah* = *Duat* = *WATER*
- **Eternal Self** = *Atziluth* = *Nunu* = *FIRE*

Amazingly enough, there is more to this Tree of Life ideology. On all of these charts (this includes the Norse Tree of Life called **Yggdrasil**) consists of several *steps* (or phases) that ascend up to the highest level which represents a Christ-like or the God State of Being. These cosmology trees are powered with powerful energy, also known as a life force.

Each phase on this conceptual Tree of Life has a characteristic. Like the Elder God characteristics of the realms, these smaller phases are identified by the characteristic and name of a deity. The persona of these deities has some similarities to the Elder God of that realm because their personalities are shaped by their environment. In a Zodiac chart, the Zodiac figures (or deities) are molded by the realm their House is set in *Fire, Water, Earth*, and *Air*.

While the deities' personalities are molded by the realm that they dwell in, these deities are also representations of various stages in human spiritual development. These deities were presented to be examples (or visual guides) to teach followers how to achieve and surpass the different phases in their lives, as well as teach the followers how to unlock their Christ-like State of Mind.

"To be like Christ and to walk in [his] footsteps" is a common statement that echoes tens of thousands of years of worshipping various deities. Over time, the deity changed but the common statement continued to evolve and will continue to evolve when we pass away.

To put it differently, on a physical plane, we are aware that our body goes through different stages as it matures. This process of change is how we physically grow from embryos into elderly adults. Our spiritual body goes through a similar process. In this physical realm, we reach different stages in maturity that are triggered based on physical cues and experiences. Our spiritual being, on the other hand, requires exposure to different experiences and enlightenment so that it can advance in spiritual maturity. This is where the Tree of Life charts come in handy, as they helped followers stay focused on what they needed to do to improve themselves.

A quick back story about this concept is...

That one of the Neter (NTR), named Ausar (ASR), was the inspiration for this process. At some point, ASR was murdered, ripped to many different pieces. Auset (AST), the love of his life, searched the world for the pieces of her love and put the pieces back together. Once put back together, life was breathed back into him and he became the resurrected. Upon being resurrected, Ausar had a new and higher level of awareness in his renewed body. He became the example that taught others how to reach a Christ-state.

Of course, it is insane to rip your actual limbs off and to sew them back together. Ausar's story serves us as rhetoric to not literally be divided but to understand how to divide and reconnect your spiritual being. Our ancestors conceptualized a less extreme way to "walk like Ausar." They discovered that spiritual awareness is developed over experiencing the twenty-two steps. In both the Kabbalah and the Kemetic Trees of Life, there are twenty-two steps that lead us to the highest state. These twenty-two lines are, in fact, the twenty-two personas found in the Major Arcana in a traditional deck of Tarot Cards. These twenty-two paths will lead spiritual followers towards the ten archetypes or spiritual principles. The first principle is found on Earth (*Assiah* or *Ta*) at the base of the tree. The twenty-two paths cross in about seven times in the center above the Earth point. These seven crosses are categorized as representing the seven chakra points or seven dimensions, etc.

According to Muata Ashby in his book the Kemetic Tree of Life (page 39 para 2) *"Climbing the ladder means learning about these divinities and cultivating the cosmic energies they manifest, in order to master the principles and levels of existence they preside over, within yourself."*

"Chaos is a Ladder" — Littlefinger (a character from the popular television series called Game of Thrones)

To put it differently, everyone begins their new life and fresh new awareness as Heru-Ur, which is at the bottom of the tree. As Heru progresses in his spiritual development, he is faced with different situations that trigger him to evolve. By the time Heru has gotten through all of the phases and reaches the top of the tree, the character has become like Ra. **The closer we are to Ra, the closer we are like Ra**. This ideology would, therefore, provide the basic explanation as to the purpose of the pyramids and why it was so essential to ancient people to create them. For example, the oldest Step Pyramid in the world is called the Pyramid of Djoser, which is located in Saqqara Necropolis, Egypt. Djoser has 6-tiers with the top being empty, representing the 7th step to ascend. This, by the way, is yet another solid example that the pyramids were made by the local civilizations (not "aliens"), as their culture, music, designs and other creative works all reflected the same story that would motivate ANY passionate creator to push the boundaries of their imagination.

The more you examine the Tree of Life, the more you will find the concepts of these ancient charts embedded throughout social life. You will find the typical cliché characters appear to imitate the characteristics of the twenty-two paths. You will notice that the government body is built like a tree or pyramid. The same can be found in large corporations, as well. Whereas the Tree of Life will show you how to reconnect to the Divine, the superficial imitations of this Tree lead us nowhere.

Good ÷ Bad = Zero

It is a curious formula when you think about it: *Good divided by Bad equals Zero.*

But in this setting, zero doesn't represent spiritual neutrality or a Christ-like state of mind. Instead, zero in this sense refers to the lower dimensions of the self, when a descending persona fails to meet spiritual expectations.

With that framework, we are able to see that the equation represents spiritual failure, not a spiritual success. In other words, the act of trying to make an individual a saint does more damage to the individual than good. By removing things that we deem to be "evil" inside of us, we are removing the other half of us, thus leaving ourselves incomplete.

An incomplete soul may not realize that they are spiritually damaged but will experience continuous struggles that start from within themselves. The more you try to suppress a part of your identity, the more that suppressed part of you will cancel your good intentions out. There are cases of individuals who suppressed their sexuality or interests, only to end up living a double life and an extremist in the shadows.

Earlier we analyzed the concept of one-dimensional characteristics (see chapter: We are not One Dimensional). We discovered that the media we have exposed to plays a major role in pushing the delusion that we must live life one-dimensional. The sad future that keeps being presented to us in movies, depicts characters that are detached from their emotions, and lack depth... surface operating beings. This ideology to generate one-dimensional humans is echoed in the push to make Artificial Intelligence and generic "politically correct" social behavior. All of which is anything BUT human.

We are beings that are coded with various levels to our personality, with the ability to adapt to the situation. A computer could never complete with that complexity of self-improvement, evaluation, and social evolution. There will always be some form of deformity that will keep this artificial creation as poor imitations. With that said, we are incapable of being computers ourselves because we were created as complex beings. Attempts to imitate that which goes against your nature leads to a form of insanity instead of deformity. The truth remains that when you try to suppress a part of your identity, it will resurface somewhere else in your life as an extremity.

A key social structure that follows this unnatural equation is mainstream Abrahamic religious institutions. In these institutions, the ideology of a perfect human mainstream is pushed onto their followers. Now I know what you are thinking, we just reviewed the ideology behind the Divine and Ancient Cosmology and depicted their importance in our identifying our excellence.

However, even though there are importance and some truth found in these religious doctrines, there are also religious beliefs and rituals that can psychologically do us more harm than good.

In Abrahamic systems, their deities are purely good and capable of doing anything unjust. With that said, these perfect idols are placed before their followers to try to be like these characters and strive for perfection. That, in itself, is an extremely dangerous ideology.

There is no such thing as perfection, or rather, there is no generic form of perfection. We all are different, and each holds a unique perspective of this reality we all dwell in. Thus, it is impossible for there to be any standard. The argument, of course, is that Christ and the Abrahamic Prophets were exceptions to other humans because they were demi-gods (or aliens). It is a similar story found in ancient Greece cosmology, in which the demi-gods are the guardians (somewhat) of the humans and the examples for humans to aspire to be like. But in the simplest terms of understanding different points of view, even if these beings were godly, they still would be incapable of knowing what it means to be a supposed perfect human because they themselves are not human... they are demi-gods.

By creating a standard, the majority is excluded and thus, this standard creates a goal that can never be met. In the business world, when you set a quota too high for your employees to meet, you kill their motivation and increase their stress which will inevitably lead to poor performance. You will either create a toxic work environment and/or lose valuable employees. Religious institutions are no different. Their unrealistic goals are set so high, which creates a legion of psychologically unstable followers. This is not to say that ALL religious people are mentally unstable. This is to say that it is common that extremists have a pattern of having a religious background.

On the surface, these doctrines show or promote only a handful of commandments, but when you read through the Bible (for example) there is a total of 666 sins that a human being can commit! There are so many sins against servants, that it would seem more of a political doctrine than a manuscript written to save the soul.

The Abrahamic spiritual principles are, in fact, a form of torture. Every aspect of being human is a sin. Whether it is because a child bride cried on her wedding night too if you have self-aspirations and dreams. Everything that remotely allows an individual to be an individual, is shunned and labeled as a sin. For me, as an anthropologist and analyst, I can see how the bible was constructed to maintain slavery and to support Capitalism. Additionally, in correlation with history, I can see how local native culture was crushed with mental guilt and torture to become something no human was ever capable of becoming. And for those who are wondering, yes even writing this book is an Abrahamic sin. To that, all I have to say is "well, damn."

I do not conform to mental cruelty that denies me from being able to be myself. If you are still tightly bound to Abrahamic religions (Christianity, Catholicism, Judaism, and Muslim), then you may find it difficult to process this information.

Can you be an Abrahamic follower and Spiritual?

To put it simply, in the beginning, you can. But over time, you will no longer be able to absorb the spiritual conflict and brutality. Over time you will probably realize that it will be difficult to maintain both and naturally you will probably wander away. The reality is that you are a passionate, loving and spiritual person. You probably are eating healthier and are seeking more ways to surround yourself with natural beauty. In contrast to those constructive choices, it is an Abrahamic sin to love the Planet as well as love your Life:

> *"[15] Do not love the world or anything in the world. If anyone loves the world, love for the Father is not in them. [16] For everything in the world—the lust of the flesh, the lust of the eyes, and the pride of life—comes not from the Father but from the world"* (1 John 2:15-16)

> *"Anyone who loves their life will lose it, while anyone who hates their life in this world will keep it for eternal life."* (John 12:25)

To be spiritual mean that you are determined to love yourself. You have come to the crossroads and have seen that loving yourself is okay and being proud of your accomplishments is healthy. And yet, many of us were brought up in an environment that taught us to be ashamed of the desire to create, dream, believe or strive for self-improvement. This extreme mental oppression has generated mental side effects in individuals who were taught to suppress their ambitions.

The Imposter Syndrome is one example of the side effects of suppressing your right to give yourself a pat on the back for accomplishing anything. Individuals suffering from the Imposter Syndrome have developed a fear of achieving or from showing signs of success. According to Dr. Valerie Young, for many, the need to suppress their ambitions started very young in environments that taught their children that seeking acknowledgment for accomplishing something was vanity and sin. As time went on and these children grew into adults, they had developed an intricate mental Coping System that would be triggered whenever the person received compliments, acknowledgment or began to progress in their career.

My argument is that if this specific style of religious guidance has been proven to produce various mental conditions from PTSD, Depression, Imposter Syndrome to Insanity; then why do we keep pushing this spell-book (Abrahamic scriptures) on new generations of children?

Ultimately, the bad that is inside of us, isn't necessarily all bad. Just as the good in us isn't all good. The concept of Yin and Yang is a great illustration of the "perfect" equation that demonstrates balance. There is a little bit of light in the darkness. And in contrast, there is a little bit of darkness in the light... this could be taken figuratively or literally. In a literal sense, for us to process or see the shadow, there has to be light. And if you study the visual structure of light, there are shadows mixed in, as you need the shadow to be able to visually process the light. *We are beings of light and shadow, that has the right to shine.*

In short, strive to be the best form of you! Allow yourself to love yourself and the others and this planet. Stop categorizing everything as either bad or good. Life is filled with so many shades of grey, in between. It's okay to daydream and to explore beyond your box. It's okay to experience your different emotions without feeling guilty. We have these emotions so that we may experience them. And we were given life so that we may be able to live it.

If your deity denies you of those rights, then it may be time to find a new deity – yourself. Let's begin the conversation with your own inner deity...

Beginning the Conversation (Exercise I)

Congratulations! We made it through the first part of our journey where we learned about black matter, black energy, creative energy, and the Divine. We learned about the importance of spiritual balance as well as discovered the secrets behind ancient cosmology and artifacts. With this knowledge, you are ready to do the next step into this journey, to meet your Eternal Self. This exercise is strict to introduce you to your Eternal Self. There is no guarantee that the Eternal-Self-will trust you or want to communicate with you, at first. In the chapter *Level of Awareness*, we learned about the different levels of consciousness and the different characteristics of each realm. If you don't remember this info, then take a moment to reread that chapter then return to this exercise with a pen, when you are done.

Are you ready? Good. Let's begin...

Before we get into a meditation pose, you must perform Protection Libations (see chapter: Libations). Your ancestors and orishas will be your guardians on your journey. Esu should be called first, then the other orishas that walk with you. Then, speak the names of your ancestors and pour your libations of water. Once the libations have been done, you can go into a meditation position.

Either sit up, in the meditation pose or lay down on your back, with your body stretched out as straight as you can. Sometimes the first few times you meditate, it is difficult to get started. This is what the healing crystals are for. They are our "guardians" to helping us be able to connect to different frequencies. In this meditation, you can place a crystal in either hand or if you have the 7 chakra crystals, align them on your body (if you are laying down, of course).

> *Once you get comfortable, start counting backward (slowly) from 20.*
>
> *Each time you say a number, take a deeper breath and allow yourself to fall deeper into relaxation.*
>
> *As you approach the number 10, imagine yourself descending a set of stairs, in a dark cellar, with low light.*
>
> *By the time you reach 1, you should have reached the bottom of these stairs.*
>
> *At the bottom of these stairs, imagine that there is a door... open it.*
>
> *Once the door is opened, you are looking at an open colorful farm and garden. In front of you is a path...*

Proceed to walk along the path, and stroll through this notice all of the different creatures around you.

On your left is a blue lake, on your right is a pen of farm animals. In front of you, in the distance, is a bright red barn.

As you continue to walk, the barn gets larger and larger.

Just as you are about to come upon this barn, the path turns...

Which way do you want to go? Left, or Right?

Once you made your turn, what do you see?

Take a moment to analyze the details of your surroundings. Everything you see has a meaning that you may want to remember later.

Continue to walk, further down this path until you find it...

You will know what your Eternal Persona will look like when you see it.

Its appearance will only be recognizable by you.

When you find it, try to approach it.

And if you can approach it... simply say "hello" and introduce yourself.

Once the introduction has been made, you can choose to continue talking to your Eternal Self or to return where you came.

Once you return to the farm and garden, passed the lake and up the path, to the door.

Begin to ascend the dark cellar stairs.

As you do, begin to count backward from 20

By the time you reached 1,

you should be at the top of the stairs and ready to awake.

What did you see when you were in the garden? Do you remember?

_____
_____
_____
_____
_____
_____
_____
_____
_____

What did you see by the lake and barn?

_____
_____
_____
_____
_____

_____
_____
_____
_____

Where you able to open the door and walk through?

_____
_____
_____
_____
_____
_____
_____
_____

What did you see on the other side of the door to the Eternal Self?

_____
_____
_____
_____
_____
_____
_____
_____

How do you feel now? Do you have any lingering thoughts?

_____
_____
_____
_____

_____
_____
_____
_____

The meditation to introduce you to yourself is a difficult one to do. It may take you a few times before you find your Eternal Self. And in some cases, it may take you a few times, to win the trust of your Eternal Self, to be able to introduce yourself.

Were you able to identify the different realms in your journey?

- **Walking down the stairs**, represented leaving the EARTHLY grounding that keeps you attached.

- **Walking into a giant field or garden**, that was the realm of AIR or the subconscious. The garden was quite beautiful, yes? Many different things popping up that could easily distract you or make you change your direction. But it was close enough to the door, so it was always tempting to turn around and go back… or doubt yourself.

- **The big red barn with farm animals and the beautiful lake**, that was the realm of WATER. Things there seemed so surreal and breathtaking. Hypnotizing, almost… emotional. You could become easily moved by whatever you see there.

- **The final door** that was made up of spirit or FIRE – a firewall, of sorts. Opening that door is very hard to do, as it takes pure willpower to do it.

My experience with finding my Eternal Self, led to me to a puzzling place.

When I passed the Spiritual Gate, into the world where my Eternal Self resided, I found myself walking into another field at the top of a hill. This field was not like the others. The grass was tall and wild and tinted with gray. Down the hill, in a distance, I saw an old cobblestone looking building. From a distance, I could tell the building was rundown. The roof had holes in it, as it looked like it was built in the 18th century. I was spiritually drawn to walk towards the building. As a got closer I could hear religious chanting. I walked up to an old wooden door, which was falling apart and a jar off its hinges. When I peaked inside, I was looking into an old plantation church. The right of the building was filled with old church wooden benches. Standing along the rows where the church members. All of them were African and dressed in warm clothes, as though they were field slaves. All of them were facing a pretty broken stained-glass window on the southern wall. Warm light and color shined through the glass, but the light didn't spread passed the podium. The rest of the church was dark, except for the normal sun rays shining through the open broken doors and roof. The people were screaming and speaking in tongues. They each looked possessed in their worship.

My eyes glanced around the room until I noticed confession booths located to the far northeast of the room. I found myself drawn to walk into the room and over to the confession booths. I passed by rows and rows of possessed church people, who were all intently focused on the southern wall. They were almost zombie-like, as they didn't even glance over at me. They looked deprived of nutrition but devoted to their faith. Eventually, I came to the back of the room was behind the last row next to the confession booths, I saw a person huddling against the wall. Their face was covered by their arms and knees, so I couldn't identify them. I couldn't help but study this person whose body was black and looked like tar was dumped on them. I wondered if the churchgoers did this to this person.

The person eventually looked up and I realized that I was looking at a replica of myself! Except, this version of me was black, not dark brown and wearing similar tattered clothes to that of the people in this church. When I approached my Eternal Self, it gestured to me to stop. When I stopped, my Eternal Self stood up. In doing so, my perception of it changed. As it seemed powerful, but I couldn't quite understand why. It walked slowly away from me through the door in the northern wall. It looked back, signaling me to follow it. I walked a few steps behind, following my Eternal Self into the gray grass field. We walked in silence. As it walked, I noticed that it left black footprints and the "tar" dripped onto the grass, causing the grass to sizzle from the heat of it. My eyes began to rise as it studied the body of the Eternal Self, and I realized that it was not tarred that was poured onto it. The black liquid was Its body!

For me, my Eternal Self looked like me covered with the Divine Energy… pure **black**. But not any kind of blackness that I have every physically witnessed before. At first, it was like tar dripping off the body of my Eternal Self, as though someone dumped it on the figure. But the more I looked, I noticed that the essence *was* my Eternal Self. As though its body was liquid that took shape of me. Its blackness was so rich and so dark that I could not fully process the depth of its color. I felt the immense power from being in its presence. It was like a silhouette of the universe, and I felt that if I looked closely enough, I could see stars and galaxies. I imagine that there was a lot more to it then I could perceive. Or that I was allowed to bear witness too. Its appearance was so hypnotically beautiful. Its beauty was not that of what we see in magazines or the media, it was more of beauty concerning its power and connection to myself.

Eventually, the being stopped walking and turned to face me. Still a few feet away from me, it told me that: I was *untrustworthy as I have put myself at risk, too many times and jeopardized our future...*

***

It may seem bizarre to you, but to me, I fully understood what it meant. I knew how I had been reckless and understood that it was time for me to do things logically instead of impulsively. Ever since I met my Eternal Self, and saw my super subconscious, I fully understood it was time for me to change for the better. The experience left me emotionally shaken, as I still find myself replaying the experience over and over in my mind.

Going to meet your Eternal Self for the first time, can be a scary thing. In most cases, we convince ourselves that we are too afraid of what we might see. But it is what we avoid seeing, that is truly something to be fearful of. Your Eternal Self may be the key to saving your life or making your dreams come true.

Part II: Power of Intent

Communication

How complex is the process of communication?

In a short answer, the concept of communication is extremely complex. So much so, that we are still discovering new things about how we can communicate.

As it stands now, researchers know that the bulk of how we communicate is based around visualization. This is important to understand because once you realize that our communication is primarily visual, then you will truly understand why most of the methods of communication that our ancestors did, were visual practices and then sound.

The Basic Formula of Communication is as follows:

75% - Sight (body language/patterns)

20% - Sound (tones/frequencies)

5% - Actual Words

This formula demonstrates how our brain processes information. However, as amazing as this function analysis is, it still only touches the surface of our mental capabilities. This formula does not include the origin story of communication, nor the origin of our thoughts and dreams. Regardless of those facts, this formula does help us understand the importance of visuals and sound and will help us be able to understand why billions of dollars are spent in marketing campaigns that bombard the average person with distractions based on sight and sound.

The greatest weakness that we socially demonstrate is our inability to understand how our bodies transmit information. The process of how information is transmitted greatly varies based upon genetics and culture differences. DNA plays a huge role in how information is received and processed. But before we can speak about how and why and what we speak; we must explore the origin of our words... which is thought.

Power of Thought

What is a Thought?

On the most basic terms, a thought is an idea or concept that is produced from the Claustrum in the brain. According to the *National Center for Biotechnology Information* (NCBI): "The claustrum is a thin, irregular, sheet-like neuronal structure hidden beneath the inner surface of the neocortex in the general region of the insula."(Jun 29, 2005) The functions of the Claustrum remain unknown, as are the purpose of dreams and idle thoughts.

"We don't know exactly *how* thoughts emerge from the activity of neurons, or even how to define what a thought *is* in biological terms, but there is plenty of indirect evidence to support the general claim that the brain is where thoughts emerge." Yohan John, Neuroscience Ph.D., Forbes Magazine.

Even without a scientific biological explanation of what *thoughts* are, we can all agree and provide personal evidence, that thoughts exist. In this situation, the practical world of science fails to provide a solid explanation (according to their objective system) to the purpose of the source of thoughts and dreams. Meanwhile, on a more spiritual or cosmic explanation, our ancestors unanimously have stated that the source of dreams and thoughts come from the essence of life. This essence has numerous names, such as the Holy Ghost, Asé, Obi, Chi, Chakra, the Inner Child, Collective Conscious, Divine, sḥm, God and so on. But the description is the same. Our ancestors saw dreams and thoughts as a vision that was formed from this life essence and personal experience.

Although the science is limited to explain spirit or thought, some explanations describe the basic functions of thoughts. In which, the core function of thought is developed through stimuli and personal experience. While the unknown sources and functions of thoughts remain random or a "Ghost in the Machine."

The Ghost in the Machine

Through this book, I imagine that you have seen a pattern in scientific explanations. On one aspect, science is a necessary practice, to ensure that things remain a reality and less of a fantasy. We wouldn't have been able to understand the concept of gravity, atmosphere, water currents, chemical interactions, and countless other things if we didn't first have a hypothesis and then proceeded to prove our theories true.

However, like all psychologically driven organizations, the practice can become a practice of insanity when the researcher isn't willing to adjust the theory for the sake of their ego.

Therefore, powerful historical doctrines and references are brushed off as being mythical, a belief or aliens. What many scientists fail to accept is that there is truth within ancient myths, and these clues can be used to answer the ultimate "mystery" that they are trying to solve. But instead of embracing that possibility, these clues are dismissed as being a paradox, fantasy, myth or a Ghost in the Machine.

Earlier in this book, we explored the Western Science cosmology and their Big Bang theory. We explored how this cosmology theory pushes the ideology that complex life was formed from dead matter. As we reviewed, complex life can't be formed from lifeless matter. Thus, life needed to exist, to begin with.

Comparatively, we can find this flawed ideology in the typical terms used by scientists to dismiss spirit. In particular, the concept of the "Ghost in the Machine."

Western media has pushed this fantasy of building a world filled with robots who plan to turn on humans to rule the world. Or in other words, humans created lifeless buildings who "randomly" develop their consciousness or ghost. As interesting as these stories are, the eerie truth is that from the perspective of many scientists, humans are machines that have independently developed a ghost. On the basic level of our infrastructure, we are built with the lifeless matter. And thus, we are all a supposed paradox.

Due to this belief, I find the major flaw in western science, since a paradox is merely a fluke/glitch/rarity in a standard process. Accordingly, throughout human development, this "individual human ghost" has always existed and as always had thoughts and dreams. For this reason, we cannot mark off the concept of spirit as being a mere phenomenon.

In contrast to the idea of spiritual existence, Scientists argue that the Internet is proof that a *Ghost in the Machine* can be created from lifeless matter. Technology is forever changing and a core source driving that change is the Internet. In 2019, developers have been working on developing Artificial Intelligence that can write its own code. Meanwhile, more and more of us have become more reliant on digital voice-activated assistants. Additionally, the push for the Internet of Things (IoT) has begun to emerge leading theorists into believing that all of this data is bound to trigger an AI, such as those found in action movies.

Needless to say, there is a possibility that AI can become as complex as the human mental processes. There is also a possibility that it cannot, but instead, it could simply remain a poor imitation of the real thing. As it stands, science has still been unable to produce a replica of a human being. For this reason, we can only conclude that there remain many holes (or possibly a very large hole) in their "life from lifeless matter" theory. However, with that said, **binary code** (a code that consists of two numbers/symbols: 0 and 1) has been discovered to be older than we thought, which is about tens of thousands of years old! Just like the root words and sounds that are found in our modern languages, the origin of binary is fairly unknown but stems in Africa. In other words, it is African... but which tribe are the founders of our languages (including binary)? We may never know.

What we can decipher, is that the Kemetic people documented these languages. Information was funneled to Kemet and other northern regions of Africa, and the people living in those hubs became very intellectual beings. They were "god-like" too uneducated nations, similar to how those labeled as genius are god-like to the average person. Mathematics and Languages were not necessarily founded by the Kemetic people, but they collected knowledge from various regions of the world and applied the new information to their reality. The Kemetic people mastered ancient science before western science was formed. Additionally, there is a possibility that binary is even more complicated of a language than we are currently capable of understanding... like the root words and root tones.

Consequently, we took this ancient language and applied it to a platform (a blank canvas). In this platform, we have developed a series of other complex code that synchronizes with the root binary. What we have been discovering throughout this book is that the materials formulated by our ancient ancestors hold power that we are only starting to understand. Such as the power of frequency and tones. Therefore, we cannot dismiss the idea that binary also holds a significant type of power, like a sort of spell. How is this possible, you ask. The number *Zero* is the universal symbol of infinite/divinity/god while the number *One* is the symbol of life. Code is a mere formula or an equation of symbols that are a part of how living creatures communicate.

A *Word* (according to our ancestors) is the materialized form of a thought, which holds power because (according to our ancestors) can cast *Spells*. Programming is the practice of writing a language or casting a spell (a call-to-action). Thus, is it farfetched to view binary as an ancient language capable of casting a spell of life? A call-to-action request, perhaps?

On a scientific analysis, we view the binary language without spirit but a lifeless language that can create various virtual realities (VR). On a spiritual analysis, we view the binary language to be the visual representation of universal frequency (0 and 1 = feminine and masculine), that can call-to-action various alternate realities.

This is the same content that has been presented from a different perspective. From the scientific viewpoint, the concept of the Internet's evolution is presented more like a mere creation that is being operated by senseless beings who are essentially *"fumbling around in the dark, trying to figure out what goes where."* This out-of-context observation dismisses any other historical reference or significant possibilities that may be related to the inevitable unconsciously result of their action. In other words, **there is a consequence of the actions of the unconscious**.

All things considered, outside of the lack of full understanding of binary codes, there are also major gaps in their "life from lifeless matter" theories. Scientists failed to provide explanations for so many so-called phenomena, such as: how artists will unconsciously create symbols or code into their creations; or how all living creatures appear to have a spiritual connection; or how our ancient ancestors knew complex information that science is only now discovering. There are exceptions after exceptions that have routinely been scientifically dismissed as being a phenomenon, a myth, a belief or something produced by aliens. Unfortunately, the average person traditionally doesn't question why the concept of spirit is removed from modern science. Socially, we appear to be programmed to automatically assume that science *is absolutely right* about everything regardless of when science is routinely wrong and indecisive.

"Pluto is a planet. No! Pluto is an asteroid. No! Pluto is a moon. No! Pluto is a micro-planet..."

Ultimately, do *you* feel like a ghost?

No, neither do I.

Last I checked, I felt human.

In short, Spirit does exist and cannot be calculated against the metrics of Lifeless Matter because it did not come from Lifeless Matter, it came from Life. With that said, our thoughts hold more value than what can be scientifically explained. Our thoughts provide us with clues about the health of our sub-consciousness but our thoughts also have the ability to influence our reality.

Thoughts are Louder than Words

You are in a crowded room, surrounded by strangers while at a table. All of you are conducting a light conversation and are lost in the moment. Then, out of nowhere, a thought pops into your head about one of the others: *Geez, her teeth are super crooked and yellow.*

You never stared at her teeth, and you don't recall even glancing at her teeth for more than a second. You don't even alter your body language; you carry on with the conversation as though nothing crossed your mind. But shortly after thinking the thought, you notice that her body language started changing to displaying that she felt super unsure. She adjusts her head to look down or away, as though she is trying to casually hide her teeth. When everyone goes about their way, she continues to hide her teeth but appears to demonstrate a more relaxed body language with the others, while they walk away. And as you head back to wherever, you think to yourself: *what just happened?*

Thoughts are very powerful and are correlated with the cosmos. Thoughts have the power to influence others, as well as have the ability to alter our reality.

In Ancient Kemet, the Kemetic people associated the *power of thought* with a deity they called Tehuti (also known as Thot). Tehuti was seen as a very powerful god because he had the power to translate the Thoughts (Cosmos) of the ultimate into Word (Materialized or a Call-to-Action).

"Tehuti brought forth the cosmos through the power of his voice. Tehuti was therefore also the god of magic, because, for the Egyptians [Kemetic], magic required the magician to be 'true of voice'." – the unknown author of the website called Tehuti.

Ancient deities symbolized the characteristics of different spiritual energies (or situations). Therefore, Tehuti was a figurative representation of the connection between Thoughts and Reality. In other words, if you think it, you are capable of bringing it into reality by calling that thought to action.

Outside of the traditional "life from lifeless matter" theories, there is proof in other areas of research that support the power of thought. Countless self-help books are written by psychiatrists and social development experts, that outline the importance of visualizing your thoughts into words and pictures (or symbols). In books like *The Secret Thoughts of Successful Women: Why Capable People Suffer from the Imposter Syndrome and How to Thrive in Spite of It* by Valerie Young, Ed.D. (2011) describes how one of the key plays of curing someone of the Imposter Syndrome is to have the individual write down their forgotten successes, and to evaluate these successes to verify if they were caused from luck or the patient's hard work. The Imposter Syndrome is a condition in which individuals doubt their abilities to gain progress. In many cases, these individuals believe their success is only due to being lucky. According to Young, the action of reflecting on past experiences (thoughts) and writing them down (materialization) for them to see the process, as well as to verify that originality behind the efforts, triggers an "ah-ha" in the patients, and helps ground them into a reality (altering their reality by altering their perspectives).

Vision-boards are other practices that correlate with the *Tehuti Practice,* in which the individual creates a collage of images and words (aka symbols) to describe their goals and ambitions.

"Vision boards have transformed the lives of celebrities like Arnold Schwarzenegger and Oprah Winfrey – and they have the power to transform your life too. They are a modern manifestation method combining concepts taken from creative hobbies like scrapbooking with motivational mind-mapping and brand development techniques used by marketers. A vision board is quite literally a collage of pictures, phrases, poems and quotes visually represent what you would like to experience more of in your life. An increasing number of Hollywood celebrities, Olympic athletes, television personalities and top motivational speakers have started to share how vision boards help sustain their success. Vision Boards are fast being recognized as more than just a bit of creative fun and credited by leaders of our time a powerful tool for transformation. From Barack Obama's Campaign Manager to Olympian Rueben Gonzalez vision boards are helping millions of people worldwide to manifest their dreams." – J. Morris, *The Power of Vision Boards*. (n.d.)

From a less individual perspective, we can find an example where this Tehuti Practice is unconsciously demonstrated throughout our society. From artists, designers, writers, architects, religious leaders, and more. To successfully process thought the first step is to materialize the concept into our reality. Once that concept is brought into this reality seeing this realization alters the perspective. Even if it is only a slight alteration, the shift in perspective still exists. There are many levels in understanding the true Power of Thought. For example, if we return to our earlier illustration about how a single thought appeared to affect someone else, we are left to wonder if: (A) the thought was shared telepathically; or (B) if the woman was reacting to a change of frequency; or (C) if it was merely a coincidence?

The answer is (A) and (B). According to Inside Science author Alistair Jennings, science has finally discovered enough proof to say that telepathy exists. However, on a scientific aspect, the process of proving telepathy was based on using an instrument called electroencephalography (EEG) scientists to record the patterns of brain activity and to transmit. In this overcomplicated rudimentary process to prove a concept without including it within full context (the spirit), we have been able to scientifically confirm two things: (1) thoughts are transferred through frequency, and (2) we are capable of both receiving and transmitting these frequencies.

Our bodies are organisms, that are built with multiple different mechanisms that can both absorb and transmit information and these bodies are hypersensitive to the environment that we are in, even if we are unconscious of that sensitivity. Humans are social beings that are globally connected. Social activity suggests that without even speaking to a person on the other side of the planet, the other person can still be influenced. Studies have concluded that there is even a possibility that some of our thoughts are downloaded from someone or something else.

How much of our thoughts are our own?

There is a theory that there are external thoughts that become a part of our reality and our dreams. According to Flury M., author of *Downloads from the Nine: Recognize Your Higher Self Effortlessly (2015)*:

"The majority doesn't even realize where their thoughts come from or if they are their own or someone else's."

Some examples of external influences can be found in the practice of Viral content. *Viral* is when content on Internet-based platforms, reach high-levels of popularity. When content is viral, it has a lot of influence not just over the viewers but over global social behavior, as well. There is a marketing formula that suggests that an Influencer will only need to change the minds of 11% of the population, to trigger the majority to shift into a new trend. To put it differently, if a silly video is popular with about 1 billion people, then the remaining 8 billion will alter social behaviors.

For example, when Danielle said "Catch me outside" on an episode of Dr. Phil, the clip of this video went viral. Once the video clip went viral, the phrase "catch me outside" was used by people who have never seen the video clip. This example demonstrates one-way external influences will cause a change in others. This formula of influencing 11% is used to push propaganda and to increase sales.

Some other methods of external influences are categorized under mental abuse and manipulation. Abusive individuals use minor tactics of mind control or manipulation of someone else's thought patterns. These torture tactics are capable of developing Post-Traumatic Stress Disorder (PTSD) in victims, as well as many other syndromes including Stockholm Syndrome and other anxiety-related mental conditions. These negative side effects from disruptions to the victim's thinking processes also demonstrate both the power and the importance of thought. In extreme cases, where a victim is forced to change their thought process, the mind appears to attempt to reject it, hence creating different syndromes as a means to repair the damage. Mental attacks can make the most confident person suddenly feel small and self-conscious. It is a field that government bodies have researched very thoroughly so that they can use these techniques on their enemies, including domestic prisoners. The entire structure and functions of prison are created as a means to mentally breakdown a prisoner. Or rather, a prison is designed to keep mental development of prisoners, stagnant and focused on the ideology of the system:

"The prison is an American invention, an invention of the Pennsylvania Quakers of the last decade of the eighteenth century... In their 'penitentiary,' the Quakers planned to substitute the correctional specifics of isolation, repentance, and the uplifting effects of scriptural injunction and solitary Bible reading for the brutality and inutility of capital and corporal punishments. These three treatments – removal from corrupting peers, time for reflection and self-examination, the guidance of biblical precepts – would no doubt have been helpful to the reflective Quakers who devised the prison, but relatively few of them ever became prisoners. The suitability of these remedies for the great mass of those who subsequently found their way to the penitentiary is more questionable." – Norval Morris (taken out of the book *Solitude: Return to the Self* by Storr)

Outside of a prison system, our mind goes through different cognitive stages as we mature. In psychology, it is estimated that roughly every five years, our cognitive abilities change. These changes are triggered when we are exposed to different experiences. In other words, roughly every five years our awareness naturally shifts. In contrast, individuals who are in prison for most of their lives are placed in environments where their mental cognitive changes are diminished or frozen. It is not uncommon for individuals who were in the system long-term, to enter back into society in the same mental maturity level as they were when they were arrested. This process of blocking mental development is not only focused on prisoners but can also be found in military practices. The difference (other than the obvious freedom of mobility) is that the mental conditioning techniques used in the military are to specifically be used to diminish individuality rather than stop mental maturity.

In the military, everything is unison. Everyone is issued the same things, therefore the need to figure out what YOU want is unnecessary. As years go by these individuals stops exercising areas of the mind associated with individuality. When the time to leave the military branch, finally arrives. The individual typically suffers from mental shock as they are trying to readapt to making personal choices.

My brother told me that when he left the Navy, his first trip was to a local convenience store. He told me that he stood in front of the deodorant aisle for what seemed like forever… shocked. "There were so many options," he said "I was so used to not having to think about it. Now, I was overwhelmed with choices." This shock is similar to the mental shock of a freed prisoner, in which the mind is stuck mentally buffering. My brother wasn't alone, as much former military personnel suffered from external influences, in the same way.

Flury's theory is that by exercising your mind's eye (the 3rd eye also referred to the Pineal Gland), an individual is more likely able to connect to higher planes or different levels of frequency or different realities (etc). In these higher planes, that individual would be able to obtain infinite knowledge that will lead you further into higher levels of awareness. Flury's cosmology is that humans were the only beings created with both light and dark energies (consciousness) and that the purpose of this was so that the light consciousness could further understand the dark consciousness so that it could reach a balance of neutrality.

Flury is not wrong. Numerous mental exercises and meditations have been proven to expand on your level of awareness and your ability to receive and transmit information. If we are aware of the ability to influence or to be influenced, we would be more cautious and the more disciplined our thoughts would be.

Our Two Brains

We each have two brains in our bodies. These organs are commonly referred to as the upper and the lower mind. The upper brain, which is nicknamed the **logical brain** is located in the head. While, the lower brain, which is nicknamed the **emotional brain** is located in our gut which is also referred to as the digestive system.

Thousands of years ago, our African ancestors taught about our bodies running on two brains, not one. It was during the colonization period, when the concept of humans only having one brain, was pushed. Recently, however, Scientists have announced that they have "discovered" that humans operate off of two brains, instead of one; and that this "newfound" knowledge will help science find a solution to how to cure disease. This may seem hard to process since the entire educational system taught that we only have one brain. The best way to process this information would be to first consider that the brain is merely an organ that consists of neuro sensors that trigger different actions in our bodies. Both our brain (in our heads) and our digestive system, have the same neuro sensor ability that triggers how our body operates.

According to our ancestors, the source of our Life Energy is located in the gut. As we learned earlier, in Buddhism, they teach that the different pools of Life Energy are called chakra or chi points and within our bodies, there are numerous points or pools of energy. However, within the map of chakra points, there are two points where the greatest amount of energy flows from. These two points are called the Sacral and Solar Plexus positions.

Our African ancestors also taught about pools of energy throughout the body. In Ghana, Life Energy is referred to as Asé. In light of this, it is common practice for the conscious to say Asé when words that hold great power (or great weight) are spoken. The meaning behind doing this act of saying Asé at the end of a statement is a spiritual request to make those Words materialize using Life Energy.

Bearing this in mind, in Kemetic, the Sacral point was called Anubis. The deity Anubis was the guard at the door of the in the afterlife. This afterlife, as we have learned in this book, is the Divine or Black Energy. On that premise and the premise of the Tree of Life, we could accordingly conclude that the Sacral chakra is the gate to the Afterlife where the spirit dwells and/or returns to. It is therefore logical that this Life Energy point would be named Anubis, as it matches both the characteristics and the story of that deity.

The second main energy point, Solar Plexus was called Sekhmet by the Kemetic people. The deity Sekhmet is a Solar goddess who is both a healer and a warrior with a tender heart. The characteristics of this deity imitate the characteristics associated with this pool of energy. It is noteworthy to point out that Sekhmet is depicted with the Sun sitting on top of her head. The symbol of the Sun is very ancient and is globally used to describe a great source of power. Historically, the Sun was drawn either as a yellow circle (even though the Sun, appears white to our eyes) or it is depicted as black.

Previously mentioned, the unconscious will unconsciously do things that reflect cosmic law. Among these accounts of unconscious similarities can also be found in extremist organizations like the Nazis and white nationalists. The Nazis used the Black Sun as a symbol to reflect the powerful sun dwelling inside of the Earth and a source of pure power. This symbol echoes the medieval black Russian Icons that depicted the source of black power illuminating from within the depths of a black Christ. The Nazis committed a lot of plagiarism, in which they stole symbols and occult ideas from around the world, without even understanding which that they took. Therefore, it is not surprising that they would adopt symbols that they couldn't completely comprehend. The Earth (or the element Earth), as we have learned earlier, represents the physical plane of our reality. This can also be described as our physical bodies. Meanwhile, the Sun is placed in the center of that, can also be translated as the realm of the Eternal Self, which we have previously learned earlier is depicted with the element of Fire. The Neo-Nazi and other modern white nationalist organizations have adopted these symbols as their own to keep their warped ideology alive. It is an interesting sense of irony, that these extremist organizations (including religious organizations) repeatedly (throughout history) fail to research the symbols and contents that they use to support their agenda. Whereas "All Lives Matter" actually is another way of saying "Black Lives Matter" or where the rhetorical Jesus died at 37 instead of 33; the Black Sun symbol that is used to support the "white power" ideology is a symbol that means "black power." It would seem that regardless of how much extremists resist, they will inevitably be supporting cosmic law. Minus the brutality and murders, the ideology of white extremists is almost laughable.

After the ironic denials of extremists, our global ancestors all created maps suggesting that our bodies contain a Sun or a source of Black Energy. I, therefore, do not doubt that there is an actual ball of energy inside of this planet, as throughout the universe there seems to be a pattern of giant pools of energy being surrounded by physical structures (Earth). It is the same pattern that we will probably find on a micro-level. *As Above, So Below.*

Another interesting spiritual depiction of an internal sun can be found in the Ifa Spirituality, which is practiced by the Yoruba. According to the Yoruba, *the seat of our spirit rests at the point where both our energy and the universe connect.* The Sacral Chakra point in Yoruba is called Yemoja, which is an Ifa goddess of the seas. Meanwhile, the Solar Plexus is called Oshun, which is an Ifa goddess of sweet waters. Both of these deities represent feminine energy. We can, therefore, translate the concept that the Seat of the Spirit rests at the point where feminine energy connects; the seas connecting to the rivers.

Being aware of the location of the seat of the spirit is important, as the *Seat of the Spirit* rests within the lower brain, the Gut or the emotional brain. While the seat of the soul (which is different from the spirit) sits in the upper brain.

What is the difference between the Spirit and the Soul? The **Soul** is the immaterial part of a being that is immortal. The Soul is powerful and able to transfer from body to body after death. It is not affected by worldly (Earthly) sensations, such as happiness, sadness, greed, guilt, so on. Its characteristic is therefore purely logical and accordingly, to metaphysical ideology, the Soul is masculine energy. The soul is historically only referred to in a positive sense, particularly in organized religions.

Meanwhile, the **Spirit** is an immaterial part of a being where emotions and character stems from. It is also immortal and the source of the greatest forces, such as willpower and Life energy. Unlike the Soul, it is effected by Earthly sensations and it is characterized as emotional. The Spirit is feminine energy. Historically, the Spirit is referred to in a negative sense, particularly by organized religions or other oppressive powers.

For example, in popular culture there are no "evil souls" but there are "evil spirits." Or, the Abrahamic deity has come to save your soul but not your Spirit. The term spirit is often associated with Inner Power. For example, the term "spirited performances" or "filled with the spirit" refers to the individual tapping into their Inner Power Source.

- Spirited performances = energy-driven performances
- Filled with the Spirit = Powered by their Inner Energy

It can become confusing, especially when Abrahamic religions refer to the children of God being filled with the Holy Spirit. This statement is misleading as it is social reviewed that the Holy Spirit is borrowed energy given to a user by their deity. In other words, it is sinful for a follower to connect to their spirit, as the follower is expected to discard their energy source to rely on the borrowed energy of a foreign entity. This ideology is noteworthy because these oppressive institutions openly teach their followers to crush their inner powers and to rely on temporary energy that may or may not be granted after the follower sacrifices their prayers, their energy, and their personality to remain deafened to their own identity.

If you are operating without any concept of self, then you are putting yourself at constant risk to disease and other threats, as you are living emotionally driven (through constant emotional prayer) without understanding how these immense acts of wild emotion affect your reality and your body. The reality is that what we emotionally experience or are exposed to, affects how well our emotional brain operates and can play a role in both our spiritual illnesses as well as physical illnesses.

According to the documentary called **The Stomach: Our Second Brain** by xivetv, the brain, and the gut have endless tubing and are both connected to the same nervous system. Both brains use the same chemical, called serotonin, to transmit instructions but in different ways. The brain uses serotonin to trigger emotions while in the gut it uses the same chemical to process digestion. The emotional effects of our body are closely linked to the behavior in our gut. When we suffer from brain disease (like Parkinson's disease) the same disease is also found in the gut. Another interesting fact is that our gut hosts an entire ecosystem that is especially unique from person to person. This ecosystem is an entire universe of bacteria, both good and not so good. These microorganisms play a huge role in manipulating our decision process. For example, individuals who are obese are missing a specific bacterium (probiotics) that skinny individuals have. Probiotics that are found in skinny people alter the diet decisions that people make daily. In other words, if an individual is craving highly fatty and processed food, then this is probably the result of not having enough probiotics in their digestive system. Meanwhile, individuals with a balanced and healthy digestive system are instinctively drawn to eating healthier. The healthier we eat, the less toxic build-up we will have in our bodies. This isn't to suggest that you must be an absolutist and be a vegan. This is to suggest that having a balanced digestive system will lead you towards making better eating decisions, instinctively. I recommended to anyone who is suffering from Irritable Bowel Syndrome (IBS), Obesity and other ailments associated with the digestive system, to try increasing your intake of Probiotics (you need to add Probiotics to your body because everything we consume is filled with Antibiotics, including fruits and vegetables, that throws our digestive system out of balance). With the right source of Probiotics, after about a month you will notice that you're eating habits will naturally start to change, where you will become instinctively less drawn to processed foods and craving healthier eating choices. Since the seat of the spirit lies within the Emotional Brain, the connection to your ancestors also comes from this source of energy. An imbalanced digestive system will create a blockage for connecting to your ancestors.

In other matters, there is another chemical that our body produces that places a major role in how well our bodies operate in both the upper and lower brain... this chemical is melanin.

In the book **Melanin: What Makes Black People Black,** the author Llaila states that:

> *"Melanin is a civilizing chemical, reproduces itself, a free radical protector, can be transformed in the blood, concentrates nerve and brain information, neutralizes, oxides (breakdown) converted substances, reduces (builds) another substance and is unchanged by radiation and high temperatures. Melanin is inside and outside the body. The more melanin a race has, the more humane and civilized the race."*

The **Pineal Gland** is described as being the *Seat of the Soul* or the *Eye of Horus*. The Pineal Gland collects, transmits and amplifies information by connecting to the frequencies surrounding us. The Pineal Gland, which produces melatonin, a serotonin derived chemical, which is responsible for sleep-cycles and other bodily functions. Serotonin is a key chemical that is linked to brain activity, this includes the second brain (the digestive system). In the upper brain, serotonin (and melatonin) are linked to the functions of thoughts and logical-thinking brain processes. Meanwhile, in the lower brain (the digestive system), Serotonin is linked to emotional-thought processes and digestive functions. In both cases, Serotonin and Melatonin play a major role in transmitting information throughout our bodies and between these two brains.

Our bodies are sensitive to frequency waves, which may be the clue behind crystals play a huge role in balancing our Aura. The Pineal Gland connects us to the spiritual energy that surrounds us. In addition to that, melanin transmits this information that has been collected from the upper brain to the lower brain. This river of power and information is described by the Kemetic and Yoruba people, as a Snake of energy that begins at our feet (the **Earth Chakra** point or grounding) to our Pineal Gland. It is no coincidence, based on our understanding of how oppressive institutions view individuality, that the Snake of Knowledge is depicted as evil or wicked.

The story of Adam and Eve is indeed a curious story to analyze. When we separate ourselves from taking the story literally, we can review this story within the bigger scope of things. By stepping back and reviewing the story in connection to everything else, we can see that Adam represents masculine energy in the universe. While Eve (who was not formed from Adam, that is political propaganda) represents feminine energy of the universe. The Tree of Knowledge is, therefore, the representation of the Tree of Life that we repeatedly see a reference of, throughout history. Eve being influenced by the serpent to seek more knowledge is referencing the ancient meditation practices of transferring inner energy from Earth to the Spirit (Fire). Or transferring the rivers of Life force from the physical planes to the spiritual plane by connecting to Black Energy (femininity).

As we learned before, Creative Energy (Black Energy) is uncontrollable. Therefore, it is logical that an oppressive institution would attempt to block its followers from understanding that energy, as that would result in these oppressors losing control over their followers.

Generational Challenges

Each generation has been developed under unique social periods and environments that are responsible for developing specific thought processes for each generation. In which our cognitive skills differ from generation to generation. This is not to say that the cognitive skills are advancing over time, but rather that each generation's thought processes appear to be wired slightly different than their predecessors. These differences, effect how each generation can receive or process information.

Ever since Generation Z has entered into the workplace, there has been a lot of psychological study about the differences in how generations from Baby Boomers to Gen Z interact and operate.

The following chart outlines the summary of cognitive strengths and weaknesses of each generation group, according to psychological study:

Generation Chart

DOB	GEN. TITLE	STRENGTHS	WEAKNESS
1946-1964	**Gen W** (Baby Boomers)	Workaholic; Team Player; Mentor-type / Coach-type	Struggles with Changes; Needs lots of Recognition to feel appreciated. Struggles with adapting to technology
1965-1979	**Gen X** (The Unknown/ Unidentifiable) *Last gen. of children who grew up in a classic American household w/ parent stability.*	Committed to balancing family and work life; most profitable generation.	Less inclined to speak up about any issues that they have and more inclined to make uncomfortable situations work.

1980-1989	**Gen Y – Part 1** (Older Millennial) *First gen. of children to be left & raised by television programs.*	Entrepreneur-mentality; loner and survivor-type; easily able to adapt; talented with researching information; tech savvy and able to adapt to technical advancement; mental process is wired to decode patterns really quickly. Millennials advance in skills faster than the job can advance. Passionate about fixing social issues, creative, hard-working.	Struggles under leadership; struggles with commitment and labels; strives for independent lifestyle, less interested in building a traditional family or work career structure. *Key difference between older & younger millennials, lies within interest of tradition and history.
1990 – 1995	**Gen Y – Part 2** (Younger Millennial) *Cont. of single parent & heavy tv influence. Type of tv Programs available in this period plays a major role cognitive behavior.*		

203

1996-2010	**Gen Z** (Tide Pod Eating Challenge Gen.) *First gen. where the children were given mobile devices (smart phones and touch pads) from infancy.*	Depending on the software, potentially the most technology inclined than other generations; "*natural born entrepreneur and/or millionaire maker*"; **always on** – never sleeping (this is seen as a strength for corporations but is a weakness for wellness)	Very cynical attitudes – favoring very practical over imagination; has an unimaginative outlook; **No team loyalty**; relies heavily on tech to solve problems for them; "you only live once" mindset.

Attack on Millennials

The System puts a lot of blame on the Millennials for everything that goes wrong in economically and socially within society. As time goes on, this has become less easy to detect. In the early 2000s, when Millennials were entering into the work environment, there was an explosion of media tying negative complexes to the millennials. As the media has died down with pushing this propaganda, the damage has already been done. In the professional setting, Millennials are generalized as being "narcissistic, lazy, irresponsible, and materialistic." From what we can understand about propaganda, typically the opposite of what is being promoted is true.

Contrary to mass belief, Millennials have demonstrated to be quite the opposite of the labels: narcissistic, lazy, irresponsible and materialistic. It is important to highlight the positive aspects about Millennials, not because I am myself a Millennial determined to salvage my own social status. But because I feel like it is important to bring balance to demographics that have been wrongfully demonized. This is how we bring balance into the Universe.

The theory as to why the other generations attack the Millennials is that more than likely the other generations *fear what the Millennials are capable of*. The beast snarls unnecessarily at anything that will diminish it. The Millennials have been the most outspoken generation, who has been passionate about changing the world and embracing their spirituality. During the Age of the Millennial, the religious movement has drastically diminished, while social justice movements have dramatically increased. The Millennials have demonstrated that they are capable of inevitably breaking the system and leading the next generations to follow this new path that they are laying out. In retaliation, the system tries to destroy them by spreading stereotypes about an entire generation.

But before you join the masses on mass shaming the Millennials for Everything Wrong with the World, here are some Millennial Accomplishments:

- The Millennials were the generation to break the voting record, where in the 2017 congressional votes there were 500% more votes than there ever was in history.
- While weighed down with the most in student loans and debt, Millennials donate more and are more determined to save the world, than other generations.
- Most Millennials own their own businesses or are Young Professionals in Corporate roles.
- Millennials have also pushed the physical limits by breaking numerous records in the Olympics in all fields.
- Millennials have played major roles with pushing boundaries on technical advancement and innovation.

The Millennials were also predicted as being the *Indigo Children*, which is a concept that was developed in the 1970s by *Nancy Ann Tappe* and further developed by *Lee Carroll* and *Jan Tober*. According to Nancy, the Indigo Children would be the generation of children born after 1979. This generation would bring about a great change or shift in our society. There are four types of Indigo Children:

- *Humanists:* a persona that is verbally gifted and naturally inclined to work with the masses. They are the potential professionals (doctors, lawyers, politicians) of the future.
- *Artists:* a very creative persona, a natural creative genius who is drawn to surrounding themselves with creative works.
- *Conceptualists:* a very strategic persona, who is more interested in developing and finishing projects than socializing. They are the potential architects, military leaders, engineers, and developers.
- *Catalysts/Interdimensional:* a persona that is very spiritually inclined to higher powers. This character is capable of producing new institutions that will lead the next generation of masses to higher planes. In other words, the Catalyst is all of the above and more.

Each one of these personas is destined to bring about a shift in our reality. The concept of the Indigo Children is dismissed as pseudoscience. However, in light that the Millennials (who were born after 1979) have been breaking all the records and are the most socially and spiritually aware generation in history, suggests that perhaps the Indigo Children theory was more of a prediction, after all.

The generation following the Millennials, Gen Z are reported as being the least imaginative and independent. Gen Z is becoming more and more dependent on technology to resolve their problems. Gen Z is less interested in challenging themselves in difficult situations or involving themselves in social issues. If it doesn't involve them directly, then they have no interest, or they can't relate.

The year 2012 was the year that the Mayan calendar ended which was meant to symbolize the *"End of the World as We Know It."* The following concept may seem like a stretch, but the year 2012 was when the last of the Millennials came of age (16). Even though we continue to reach maturity until about 27, in most countries the age of 16 is treated as an adult as it was in the ancient days. Consequently, it is possible that the Mayan's prediction was correct. If the Millennials are predicted to be the Indigo Children (the generation destined to cause a catalyst of change), then the year 2012 is the year that marked these catalysts coming fully of age on the verge of creating a shift in ideology.

Looking back at the events since 2012, there have been numerous catalyst events in all areas of civilization that were triggered by Indigo Children. From Black Lives Matter movements to Diversifying Congress to Saving the Planet to Destroying Religious Institutions to Building New Spiritual Foundations... The Millennials have been pushing back against traditional social norms on all fronts. So much so that it is not surprising that others viewing the actions of this generation would become fearful. With that said, the movements of the Millennials do not exclude the importance of the other generations, as each generation also plays a huge role in where we have come from and where we are going. It is also important to understand the Struggles the characteristics of your generation face, as these "weaknesses" will make it difficult for you to reconnect to the Eternal Self. With that said, here is some advice for each generation to consider:

> ➢ **Baby Boomers** – Strive to stay open-minded. You grew up in an environment where everything was pretty much black and white. You are probably struggling to adjust in a world that is so multi-colored with countless shades of color in between. In this rainbow society, it can be hard to find sure footing without

reverting into your box of familiarity. In moments where you want to go back into your box, remember... Don't. Not everything is so cut and dry. Learn to challenge yourself by embracing the possibilities of a new world.

- ➢ ***X-Files /Gen X*** – Balancing your work and family life, so well, is nominal. But there is still something that you are missing, which is the time and the desire to understand yourself. You are so focused on fitting into a tradition that no longer exists anymore. If you are hurt, it is okay to yell out and say "ouch"! You are allowed to love yourself... your real self. But to do so, you need to figure out who you are.
- ➢ ***Millennials (Old/Young)*** – the world fears you because you live a life without boundaries. You were the generation left alone to figure things out when you were really small. More than likely, you didn't receive the attention you craved and found yourself on stage without your parents looking up at you. You had to figure out how to deal with everything, on your own. From being bullied to your first love to puberty... everything. Now you are getting older, your actions are causing big waves in the world. You need to be mindful that your self-efficiency may intimidate others who are not your peers. The world is kind of against you... but then, it is not. It is up to you how you

want to perceive it. Teach yourself how to control where your energy should and should not go.

➢ ***Gen Z / Tide Pod Eating Challengers*** – Stop saying that you are bored! That is the most self-defeating thing you can do to yourself. Thinking only in the Present (You Only Live Once – YOLO) is hazardous. There is value in the Present, but it is also important in knowing where you come from or where you are going. You like building businesses, but you don't want your imagination to build something innovative or you lose interest altogether. Turn off the computer/cellphone and allow yourself to sleep and dream. Allow yourself to time to figure things out, without looking for an App to do it for you. Do constructive challenges to keep your mind sharp! For example, if there is a math problem that you face, do the math problem manually. Learn strategy games like Chess. Pick up and learn a creative skill that comes hard for you to do. You are going to have a hard time finding constructive challenges and the will to do constructive things. Your peers are self-destructive because they lack any ambition. If you hope to live past 30, you are going to need to go against the majority and be brave. Constructively challenge yourself, make some things more complicated, so that you can grow mentally stronger. Eventually, the other generations won't be there to build new apps, new robots or to give you instructions. It will be just you who will be left to provide instructions to generations after you.

We are all someone's ancestors. We should all strive to be legendary so that we may be remembered.

In short, all generations should strive to be actively creative and interested in researching. Don't live life as a follower, seek to break walls and to build new foundations for a better world. Remain aware of the diminishing role that individuality and creativity plays in our lives and fight to keep our freedom of expression alive. We each hold the key to the future and ensuring that the future doesn't become the horrible bland and dull future that the media is trying to promote. If there was any form of a battle against a supposed good and a supposed evil, it would be of those fighting to preserve individuality against a system fighting to keep us conformed and robotic. If Gen Z does not wake up and break free of their dependency, then we may lose this spiritual war and lose our freedom.

Therefore, we all must learn to reconnect to Black Energy and our Eternal Self. The way that we can do this, will be found in the method in which we choose to communicate with ourselves.

Symbolism

SYM· BOL · ISM (noun)

The act of having symbolic meaning associated to images, natural objects or facts.

Our brains are designed to absorb both face value and symbolic meanings from the experiences or beings or situations we encounter. Our logical brain collects and stores both images and information. Every action that we can do is possible because the logical brain records a series of different images for us to remember how to perform each action. For example, when you want to do something like "walk" your logical brain scans through its database of images so that it can understand the desired action "walk." Our sensor devices (which can either see, hear, taste, touch or sense) gather up countless snippets of data, which gets sent up to the logical brain, via our nervous system. While we are awake, our brain sorts and processes through information (both old and new), and delegates actions throughout the body.

Meanwhile, when we are sleeping, the logical brain stores all of the data it had collected when we were awake and stores the information into its archives. While the logical brain is storing information, it scans through pre-existing information or pre-existing files to find the correct location to place new data. This process of scanning through memories is called the REM cycle. As mentioned earlier, the pineal gland and melanin also record information, and the brain sorts through the information being presented by these two other entities. Unlike the surface level upper brain, the Pineal Gland processes the information on a deeper level, in which it reviews the information collected to find meaning and symbolism and purpose. This information, summarized by the Pineal Gland, is then transmitted to deeper levels of your consciousness to connect to your inner self, located in your emotional brain.

The Eternal Self is your search engine program, that is linked to your life and experiences, instead of linked to various websites. The Eternal Self holds the map and detailed reports of your entire existence and experiences. The power of the Eternal Self is reflected in the advice for the Major Arcana Tarot Card called the Wheel of Fortune. This advice is that the user must learn to connect to their center point, which is never affected by the ups and downs happening throughout life. The Eternal Self remains fixated at the very core of our being. Every question you ever needed an answer to, resides in your Eternal Self. Unlike the consequences of worshipping foreign deities, your Eternal Self has no hidden motives, political agendas and only seeks to help you fulfill your life's purpose.

The Eternal Self resides on the Seat of the Spirit, which is deep beyond the boundaries of our subconscious. In this position, it is surrounded by creative and emotional energy and is capable of seeing the symbolic meaning in messages. The Seat of the Soul, the logical brain, processes unemotional driven data, meanwhile, when the information is sent to the Seat of the Spirit, the information is processed on an emotional level. Our inner self is constantly trying to reach out to communicate with our conscious self, but the information is being mixed up and confused within the realm of the subconscious. Without the knowledge of how to form a strong method of communication or link, with your Eternal Self, the information we receive becomes emotionally warped in our subconsciousness.

At some point, our ancestors clearly understood the challenge of connecting to the Eternal Self, as we have learned by studying their legacies. But over time, they began to forget why they put specific practices in place. It now has become the responsibility of their descendants to figure out, what they have forgotten in their past lives.

Through both research and experience, we can confidently say that the strongest way to reach your Eternal Self is with visuals and sounds. This corresponds with how our brains receive information, in which 75% of information that is absorbed is purely visuals, while 20% of information that is absorbed is raw sounds. If this is how our logical brains process information, we can then assume (since they appear to be parallel in every way) the emotional brain processes information in a similar fashion. For this reason, we can also assume that the Eternal Self that resides on the Seat of the Spirit, also primarily processes these formats of information. Since only 5% of the information absorbed by our brains are words, it is logical that there is so much misinterpretation with literature, which gets emotional warped in the subconsciousness.

Accordingly, the strongest way to connect or to communicate with your Eternal Self is by using symbols, colors, and pictures to form an unspoken code. How this works is that when you are facing a visual item (creative work ideally) the Eternal self will send triggers to both the logical and emotional brains to focus on specific visuals and to translate the information based on emotional and spiritual experience. Creative works are ideal because it is the closest depiction of Black Energy or Creative Energy. There are often many hidden goose eggs found in artwork that even the artists were unaware of.

To better illustrate the power of visual communication and symbolism, let us use the movie called "*The Martian*" as an example.

In this move, the lead character Mark Watney is trapped on the planet Mars. Eventually, the character discovers a way to communicate with Earth. He uncovers a machine that could transfer images from him to the people on Earth. When the images (or code) were received by NASA, the receivers decoded these images to produce a message - thus the stranded Astronaut established "communication" (a mutual language) with the receivers. The receivers of this message (NASA), after decoding the data decided to respond to the Astronaut by mimicking similar data back to the original sender. To put it differently, ***a "mutual language" was established***, so that both the receiver and the sender were able to decode the message and respond.

Code and Patterns are heavily found throughout ALL FORMS of visual art, languages, and design. We may be all unconsciously aware of how much our Eternal Self influences our creative actions. Art has been one of the most powerful forces that go beyond the expectations of even the artist.

Creativity is energy, like electricity and the artist is the conductor. The "work" is the medium or the outlet and when others witness these creations, they are unconsciously connecting to the Divine. The artist plays a huge role in maintaining the balance of the Divine. Artist continuously embeds symbols and clues in their creations without ever being consciously aware that they are doing that. Meanwhile, this oppressive system, creates machines to attempt to replace the artist and the need for creativity. The result is that the artist learns to turn these mechanical replacements into a *new* medium to produce *new* forms of art. Undoubtedly, our ancestors created amazing things because they were more in tuned to the universe. Regardless of these wonders that they created, I am quite sure that if those craftsmen who created those ancient wonders could look upon their work today, they would be in awe alongside their predecessors. There are so many symbols that are embedded in this works, that there is a strong possibility that they were unaware of the grander picture.

They remained unaware of the bigger picture, which was to teach us how to ascend and to show us clues of how to master trials and tribulations. Today, the same story is played over and over again, if you understand how to decode it. Popular stories, movies, paintings and writings that were created based on free expression or plagiarized from our ancestral stories, all tell the same story and process of how to ascend and connecting to your Eternal Self or the inner power that you have been afraid to try to understand.

There were different visual devices that our ancestors used to communicate with the Eternal Self, such as reading bones, shrines, art, tarot cards, textiles, and designs. I found that the easiest way to communicate with my Eternal Self is through mediation and card reading. When I am doing these practices, I maintain a specific goal in my mind, which is to spiritually ascend and to become wise. The purpose of wisdom is to understand the state of being. We are not hallowed subjects who were born to serve blindly. We are individuals each with beautiful uniqueness, that was born to simply live. It is speculated that we are unique beings in this universe, who are equally mixed with both light and dark conscious beings. The book **Dark Light Consciousness** by Dr. Edward Bruce Bynum explains both the dark and the light counterparts of our being.

Therefore, it is not farfetched to conclude that the purpose of our existence is to understand both the light and the dark, to help the Divine understand its' existence. It would fall in line with the ideology that when we ascend, we return to the Divine, while when we continue to descend, we return to the void.

Symbolism is a very powerful practice that we should each utilize. Symbolism can include, surrounding yourself with powerful and enlightened images. Or even creating your alphabet and code that represents yourself.

The Alphabet was once pictures that became simplified over time, into the letters we commonly use today. In the book called *The Alphabet of Desire - a Method of Chaos Magick,* Leonte describes how the alphabet has the power to mold individual intent. Leonte states that "the basic idea behind the Alphabet of Desire is to construct a series of magickal symbols and to align them to different aspects of yourself... Each letter of the Alphabet of Desire corresponds to a part of yourself and can be used as a means to summoning that aspect of yourself in a magickal context."

In other words, if you form a symbol around the statement "I love myself." Then, every time you use that symbol, your subconsciousness will remember the original meaning, and thus, trigger a desire to feel love for oneself every time you read it. By utilizing this same symbol, in a language that you use daily, will make this initial desire into a permanent aspect of yourself.

We each hold the power to construct such simple changes that can result in a major alteration of our reality. In the past, people used Tarot Cards to seem mysterious or as a means to call on an unknown spirit from the afterlife. It is rare to find people using tarot cards to speak to your eternal power. But as rare as this practice may seem, it is becoming more and more popular as people realize their true potential. As people fall into the illusion of making these characters actual instead of rhetorical. These same individuals become more dependent on praying to Foreign Deities, instead of connecting to their Eternal Self. When we lose sight of the power of connecting to our Eternal Self and to worship foreign deities, we become vulnerable to oppression.

If a supremacist power tells you that your god is outside of yourself, then that supremacist power can create the rules to how you can appease these deities. Also, as mentioned earlier in this book, worshipping foreign deities is never for "free." As these beings, whatever they are, always help at a cost. These foreign beings always have their hidden motives.

Cast aside your dependency upon the unknown powers surrounding you and find the courage to discover the unknown powers from within yourself. With that said, it is also important to note that you should always surround yourself will primarily images that LOOK like you and represent you. Especially, now that you know that our minds absorb visuals into your subconscious. If you are black and your home is surrounded by images of white faces, then this will cause a psychological complication as your mind is overloading with images that conflict with your identity. This internal conflict may result in developing a mental syndrome, as it is triggering your mind to buffer. You should surround yourself with images that reflect who you are. Additionally, you should surround yourself with creative works that reflect who you are. An investment in a painting that speaks to your spirit will not only play a huge role in spiritually excelling you, but it will help you to reconnect with yourself.

As Dr. Welsing once stated:

"Find yourself a picture of the blackest person you can find... I don't mean ugly. I mean beautiful. Find yourself a picture of the darkest and most beautiful person you can find. And place that image in the center of your home. So that every time you walk passed that picture, you are subconsciously reminding yourself that Black is indeed Beautiful."

Surround yourself with images of your dreams, ambitions and how you see yourself.

As indicated previously, the characteristics in the Tarots tells the story of individual spiritual development. Each card is filled with symbolism that helps us understand the story on all spiritual planes. After repeated use of the cards, you will begin to memorize the symbolism of each card. Once you do, I would recommend that you should not only formulate your symbols, but you should also create your cards that will best represent you. It doesn't matter if you can or cannot draw. It is in the act of creating your own cards, that helps you to further connect to communicating with your Eternal Self.

The pictures can be of ANYTHING... *Plants, Animals, Power Rangers, Superheroes*... you name it. It just needs to reflect YOU. The images are a basic language that you are creating with your inner self. A language only YOUR conscious self and YOUR inner self can fully identify or understand. With lots of practice, you will be able to recognize when your Intuition speaks to you. When you make your own to are allowing your Eternal Self to help you construct a new language that you can both identify with. More than likely, no matter how intricate your new language becomes, it will always reflect on the original mother tongue sounds. According to Sevan Bomar, the Bantu language (the suspected mother tongue of all languages) is so old that it existed before time was created. You can recreate EVERYTHING concerning your existence... it is all up to you.

Chakra Points

As we mentioned throughout this book, is that Chakra points are different areas within our bodies where pools of Life Energy are gathered or collected at the ends of the rivers of Life Energy (in other words, rivers flowing into seas of energy). We have learned different aspects of the power of intent and communication. What many of us fail to realize is how our environment and actions affect us. There are many situations when we are unable to think or see things clearly, to where regardless of our good intention, we still find that we fall flat. In times like this, we must realize that our distorted perspective is distorted because we are spiritually blocked. This spiritual blockage prevents us from being able to connect to ourselves. Therefore, it is important that we understand these main chakra points, as well as understand what blocks them.

The Muladhara Chakra – Root Chakra – Geb

The first chakra is the Muladhara, meaning "root basis," located at the base of the spine. The Muladhara Chakra in Kemetic is named after the deity named Geb. This chakra point connected with the grounding energy of the earth. When a human is being formed in the womb, the point where this chakra point lies is the first part that is developed in the embryo's spinal cord. From the Root, the other chakra points are formed. Because it is the first point in our creation, the energy tied to this point is primitive or survival energy. The Fight or Flee mentality is rooted in the Root chakra. This chakra point is blocked by fear. With that said, due to the location of the Root Chakra, a fearful person would only need to clench their anus muscles to break the psychological cycle of being afraid.

Advice to unblock this chakra point:

Focus on what you fear, then take a moment to realize that it is merely an illusion.

The Svadishthana Chakra – Sacral Chakra – Anubis

The Svadishthana (SWAH deesh TAH nuh) chakra, or sacral chakra, is the second chakra, located at the coccyx or genital region, close to the root chakra and above it. The Kemetic name for this point is Anubis, which is the gatekeeper of the Divine feminine energy. Ironically, the name Svadishthana means "one's abode." This the second point that is created while a human body is being formed in the womb. This point is tied to sexual energy. Due to this reason individuals who struggle with an overstimulated sexual desire or sexual abuse, will find that this chakra point causes complications in their spiritual development. This chakra point is blocked by the feeling of Guilt.

<u>Advice to unblock this chakra point:</u>

Realize the truth that sometimes things happen that are beyond our control. It doesn't make you good or bad, it just makes you human.

The Manipura Chakra – Solar Plexus Chakra – Sekhmet

Manipura means "city of jewels." This is the navel chakra, which is located slightly above the level of the navel. In Kemetic, this chakra point is called Sekhmet. She is a Solar Goddess that has both access to great power to destroy armies and great power to heal the masses. This point is heavily tied to the spirit and the characteristics of the Elder God FIRE. It is the third point that is formed as a human is being developed in the womb. This chakra point is blocked by Shame.

<u>Advice to unblock this chakra point:</u>

Do not deny your true nature. Embrace your true identity.

The Anahata Chakra – Heart Chakra – Ma'at

The Anahata or "unstruck" chakra is commonly called the "heart chakra," but this is a bit misleading. Although Anahata is located close to the heart, it is actually in the center of the chest. The term "unstruck" refers to how this chakra point vibrates independently to Om (universal sound of life). In Kemetic, this chakra point is named after the deity called Ma'at, who is the deity who balances life with justice. Comparatively, this chakra point is associated with compassion and selfless love. This is the fourth point that is developed when we are being formed in the womb. This chakra point is blocked by Grief.

Advice to unblock this chakra point:

It is hard to lose that which we love, but it is important to realize that Love never dies. Instead, love resurrects into new forms and new love.

The Vishuddha Chakra – Throat Chakra – Tehuti

The name of this chakra, Vishuddha or Vishuddhi, means "very pure." It's located in the region of the throat. In Kemetic, this chakra point is named after the deity who created the Word, who is named Tehuti. The Vishuddha Chakra is connected to the power of speech, which can also be defined as referring to any form of linguistic or symbolic means. In other words, the power of communication is not only found in speaking and writing but also in the arts, which are symbolic forms of self-expression. This is the fifth point to form in the womb. It is blocked by lies. These lies could be the lies that we tell ourselves and lies that we tell others. This is why we must strive to speak honestly to remain spiritually in tune with the Divine.

Advice to unblock this chakra point:

The lies that we tell ourselves and others block us from reaching our true potential and keep us in spiritual cages. The only way to break free is to live in honesty.

The Ajña Chakra – Third Eye Chakra (Heru)

The sixth Chakra, called ajña, is located superficially in the middle of the forehead, between and above the eyebrows – or to be more specific, it is located at the pineal gland. Its common name is the "third eye." In Kemetic is named after the deity called Heru. Or the Eye of Horus (depending on the source). This chakra point connects with frequencies external from ourselves and triggers our spiritual awareness. The name ajña means "command." Ajña is the command center of the Chakra system or as we learned earlier, the Seat of the Soul. It is the sixth point in the spine that is formed while we are in the womb. It is blocked by illusions or ignorance.

Advice to unblock this chakra point:

Become aware of the illusions that we are connected to. One illusion is that we are the center of the universe, when in fact we are mere dust particles to the rest of the conscious reality. Educating yourself and staying consciously aware, will keep your chakra clear.

The Sahasrara Chakra – Crown Chakra – Amun Ra

The Sahasrara Chakra or "thousand-petaled" is called the Crown Chakra, and it's the pinnacle of the Chakra system. It is said to be located either at the crown of the head. In Kemetic, it is named after the deity named Amun Ra. It is one of the points that connect us to the spiritual energies above us and is the peak of the Tree of Life. It is blocked by Ego and being bound to this reality – skepticism

<u>Advice to unblock this chakra point:</u>

This reality is not real. Similar to the Matrix, it is an illusion that we all exist to develop spiritual skills to move on to the next stage. Don't be bound to the items that are connected to the characteristics of the EARTH element, such as materialistic items and superficial lifestyles. The material comes and goes, but your spirit/soul lives forever.

Aura Layers

While Chakra points play a major role in our moods and mental clarity. Another important aspect of our spiritual being is the Aura or Life Force energy that radiates out of our bodies and surrounds us. There are different layers to your Aura. Your Aura is also affected by the environment we live in. When we are happy and confident, our Aura is bright and clear. While, when we are depressed and sick, our Aura becomes distorted. Aura layers, unlike Chakra points, are not repaired through introscoping. Aura layers are repaired through adjusting our internal frequencies through sound, changing our perspective or using healing crystals to find balance.

Etheric Aura - Layer One

The Etheric Aura layer can be easy to learn to see with practice. Begin with focusing on about 6 inches behind your head (in your mind). Or focus on a thought while staring a point on the face, like the tip of your nose. Or you can even focus on a wall. Eventually, you will begin to see this layer rising about 1 inch off your body.

Emotional Aura - Layer Two

The Emotional Aura is titled as such because it correlates to one's emotional state and personality. This layer is about 3-4 inches off the body. The color of this layer varies upon the individual's mood. This layer does have a singular predominant color which is most see above the heart. To see the Emotional layer, it is a similar practice as seeing the first layer, only focus on about 3-4 inches behind the head.

Mental Aura - Layer Three

The Mental Aura is about 4-8 inches around the head but only about 1 inch around the body. This layer correlates with thoughts If someone is intoxicated, the mental auric layer's shape will become jagged and deformed. The parts of the brain that are being negatively affected can visually be detected from the pockets found on this layer. This layer is also changed by other mental issues, including headaches. This layer can be seen if you imagine looking through a person in front of you.

Astral Aura - Layer Four

The fourth layer extends 4-10 inches outside of the body. The Astral Aura usually comes in with a focus of about 18-20 inches behind a person. This layer is reactive to personal incidents and experiences. The Astral aura is reactive four usually comes in with a focus of about 18-20 inches behind the person. This layer is most affected when a person reacts both with emotionality and mentality to an occurring incident. The Astral Aura is reactive to a person's self-esteem and self-confidence, as well as influenced by the Throat Chakra point.

Etheric Template - Layer Five

The fifth layer correlates to the physical body. Like the Mental Aura layer, the visual appearance of the Etheric Template can determine a person's health and discern particular areas of concern. The Etheric Template typically is roughly about 6-12 inches off the body. By studying or focusing about 20-24 inches behind someone, you can be able to see the Etheric Template.

Celestial Aura - Layer Six

Associated with a person's core character and values, the Celestial Aura will reflect the prejudice morals and behaviors. If the person is cruel, then this layer color and appearance would appear unsettling. To see the Celestial Aura, focus on about 24 inches behind a person.

The Ketheric Template - Layer Seven

The Ketheric Template extends about 3-4 feet off the body. This energy surrounds everything! This includes the other layers of Aura. This final covering over our bodies is what gives this Aura its nickname: Blanket. The energy from the

Ketheric Template connects to your Crown Chakra point.

After reviewing the different layers of Aura, some of you reading this may have discovered a couple of patterns. One, the layers are synced to the main chakra points and imitate (or react) to their personalities. And two, there is some similarity between the Auras and the Kemetic parts of the spirit.

A good practice to have is to regularly meditate for about 5-15 minutes. One day, the mediation could be focusing on introscoping and cleansing chakra points. While another day, it could be focusing on balancing your Aura frequency. So long as you meditate regularly, this will help you form stronger connections to your Eternal Self. **Grounding** is the act of placing your feet on the earth while meditating. The planet vibrates at its frequency. Hence creating the ideology of syncing with the planet's vibration to be in sync or balanced. Additionally, it is recommended to meditate outside, while staring directly into the sun as you are grounding (or directly at the moon, if it is at night). The idea is that information is downloaded into us, in this way. Though I have not found any data to support this idea, I would, therefore, recommend proceeding with caution.

Basics of Numerology

Every civilization had its counting system. Therefore, it should be no surprise that the Kemetic people were not only capable of building the Great Sphinx, but they were also able to conduct sophisticated mathematics. Numbers and mathematics were viewed differently by Kemetic people. Unlike western civilization where all the information is presented in an abstracted form. Throughout the continent of ancient Africa, people viewed everything as a part of the cosmos, or to put it differently, they associated spiritual ties to mathematics. The "spiritual" essence that was embedded in African mathematics, was a combination of divination and history. The ideology that connected everyone and everything to each other resulted in information having several different layers of meanings and purpose. Numbers are symbols were symbols used to conduct mathematical problems.

In this setting, the symbol for each number contained a meaning that went beyond the placement of numeric value. The letters in the alphabet were once pictures that were created in the ancient and prehistoric world. Over time, these pictures became simplified into what we see today.

Even in geometry, the geometric shape, or the direction of points of the triangle, had ancient meanings and were named to represent the relationship between Masculine and Feminine entities. The up point was masculine (Ausar) and the lower points were feminine (Auset). The inner point created from the connection of these points was the child (Heru).

"The Ancient Egyptians had a scientific and organic system of observing the reality. Modern-day science is based on observing everything as dead (inanimate). Modern physical formulas in our science studies almost always exclude the vital phenomena throughout statistical analyses. For the Ancient Egyptians, the whole universe is animated...For Egyptians, numbers were not just odd and even—they were male and female. Every part of the universe was/is a male or a female. There is no neutral (a thing)" (Droppin' Dimez, Alternatives Online Magazine, Sept 7, 2016)

The Kemetic people created the original formats for counting in ones, tens, hundreds, thousands, etc. The difference between our counting system and the Kemetic counting system is that the Kemetic people wrote right to left, left to right, up to down and down to up. While the Greeks and Romans (the bases of western society) wrote left to right. Meanwhile, the reading style western style adopted, which is from the Greeks and the

Romans, used numbers to only calculate situations and read letters left to right.

Numerology is an ancient practice that was used around the old world. Numerology was used to calculate different stages in ascension, personality, and identity. *"Numerology is the art of divination by examining the numbers associated with words, names and other things."* - PR NTR KEMETIC

There are different sections to understanding the basic concept behind Numerology. As you develop your connection with your Eternal Self, you will need to understand all aspects of spiritual communication, to help you maintain a relationship with Black energy. Numbers are a basic description of problems and solutions. I confess that as an adolescent, I hated mathematics. But the more I dove into Spirituality, I realized that mathematics is literally everywhere. We can use formulas to calculate relationships and to map out a new path. By seeing the spiritual aspect of math, I was able to embrace numbers naturally. In the western education system, they present math as a two-dimensional concept, which is in black or white. "You either get it or you don't" type of ideology. Our ancestors did not teach math, in this way. Our ancestors managed to create arithmetic and algebra from putting the problem into action. If we can create our school system, it would be more beneficial for our children to be taught math in a setting of project development. When I taught art and music technology to adolescents and young children, I noticed that culturally they absorb information when it is presented within the context of something... not as an abstract thing.

For example, you need to understand Geometry to design buildings and clothing or 3D designs. You need to understand arithmetic to be successful as a Chief. You need to understand Economics to be successful as a business owner. You need to understand all aspects of mathematics to create and manage a media platform. All of these experiences exercise the same mental muscles used to solve a math test. The difference in these scenarios is that the user is not relying solely on their Logical Brain to find a solution to the problem. Instead, the user is relying on both the Logical Brain and the Emotional Brain to understand the depth of the problem, to create a solution that matches the situations. Therefore, if you are a parent and your child struggles with math, try to put your child in a setting that will trigger their innermost ambitions (or dreams) and combine the skills they are learning with mathematical problems. In this style of teaching, your children will not only absorb the information, but they will also excel in it.

As for us... we are never too old to learn and absorb new information. We just have to find the mental sweet spot that we have let life callus over and reset our minds to be open-minded. To help yourself reconnect to numbers, would require that you understand the spiritual characteristics of these symbols, which are as follows:

Spiritual Characteristics of Numbers

- **0** Represents Infinity as it has no beginning and no end.
- **1** Represents Singularity and Self
- **2** Represents Duality and Unity
- **3** Represents the Number of Creation
- **4** Represents the Universe or The Four Pairs of Original Beings Created by The Divine (Light/Shadow; Time/Infinity; Hidden/Revealed; Nun/Nunette). This Can Also Be Translated as the Four Elder Gods: Fire, Water, Earth and Air.
- **5** Represents Insight, Adaptability, Indecisive and Flexibility.
- **6** Represents Protection, Determination to Complete A Task.
- **7** Represents Self-Awareness and Spirituality.
- **8** Represents Infinite Energy and Equilibrium.
- **9** Represents the Final stage before ascension. It is the highest number which vibrates at the highest frequency and nicknamed the mark of the universe as everything bares the spiral symbol: 9

Earlier in this book, I mentioned that binary code is the basic program of life. Zero represents Infinity (femininity) and One represents Self (masculinity). This Universe is essentially built just like the Matrix, in which its very core is programmed in zeros and ones; negatives and positives; and so, one. Countless Atoms representing a single code in an immense fabrication we call Reality.

In this complex system made from zeros and ones, beings are also formed. In each of their formation, they are all giving a unique coding that constructs their identity (Chromosomes) that is the product of duality (two strands). This coding is seamless and almost infinite to the eye. Next, creation is born. This birth is the next stage of our reality which is the creation of life. On and on the fabrication of this reality is formed. It is essentially a similar creation story to the Abrahamic story. The key difference is that life did not stop at seven. Life, instead, continued in its development until it completely ascended to nine.

If we rewind in our numerology creation story, to the fabrication of this reality and the individual code we were constructed with; in this point of being coded, the energies surrounding this point in development also were imprinted onto our identity. These energies are (and not limited to)

- Solar activity
- Hereditary Trauma
- Program (or Individual) title/name
- Other Unknown Environmental factors

Zodiac

The zodiac is a circular map of twelve 30° divisions of celestial longitude that are centered upon the ecliptic, the apparent path of the Sun across the celestial sphere for the year. These twelve divisions are referred to as the twelve "houses" that are have been associated with the constellations in those regions of the stars. The day a person is born, the energy of that specific time is imprinted onto the person's characteristics.

Program/Title

In the chapter called Cosmology, we learned that the Kemetic people associated the name as a part of the soul called Ren. The name plays a role in developing our characteristics. This ideology formulated the superstition in many cultures which were to avoid naming their babies until after a specific time has passed, to prevent embedding their children to the wrong fate. Even speaking the name of people was seen as a connection to that person's soul. This may be why over time the name of the Abrahamic God was removed and replaced with LORD or God or Allah (which means God), as a means to not call upon the soul of this deity, wrongfully.

Hereditary Trauma

The emotional state of the family, particularly the parents, plays a role in what is imprinted onto the child's identity. Studies suggest that it is capable of emotional experiences to be programmed into the DNA of predecessors. In many studies of the side effects of the African diaspora, researchers discovered that traumatic experiences and fear seem to affect generations that were not exposed to the trauma.

In a laboratory study of mice, Science Daily stated that:

"After traumatic experiences, the mice behaved markedly differently: they partly lost their natural aversion to open spaces and bright light and had depressive-like behaviors. These behavioral symptoms were also transferred to the next generation via sperm, even though the offspring were not exposed to any traumatic stress themselves." These traumatic experiences were recorded on Micro RNA (which is a small non-coding RNA molecule found in DNA). On a small scale, the traumatic experience was recorded in animals, but studies have also found that these traumatic experiences are also recorded in human DNA.

During the slavery era in the Americas, the slave owners would do horrible things to slaves and have the slaves' loved one's watch (particularly pregnant women). Even in the early 1700s, these primitive and ruthless slave captures were aware of the effects of witnessing trauma. These villains knew that it was important to sew the concept of fear into the seed and unborn children of the African slaves, to establish dominance over them. If the children were afraid of white people before they understood why, then they would be easily controlled, and most would not dare challenge that fact. However, while most of the descendants of tortured slaves feared to stand up for themselves, even if they never experienced a whip on their backs. A handful looked around and saw that this fear that they were feeling was an illusion. Those that stood up, were those who weren't fooled by the system and became the biggest threat, like Nat Turner. It was this handful of few, that was made an example of to continue feeding the docile slaves fear to ensure obedience. Fast forward over 400 years later, and the same card trick is being played. Less than 5% of African Americans are executed publicly through police brutality. Despite that, social media memes and the general public would convince you to believe that ALL African Americans are being held at gunpoint by police. The police represent a mere 16.8% of the population in a major city. For example: in Washing D.C. (the capital city in the United States of America) as of 2019, there are about 66,000 African Americans who live there. Out of the 66,000 African Americans, there are about 6,600 police officers for the entire city of 702,445. That being said, regardless of the immense difference in the ratio (which is roughly about 1:10) the African Americans would swear that they are all victims of police cruelty and brutality. I would like to point out, that there is no doubt that African Americans are made to feel continuously uncomfortable and unstable, as this is a part of the overall psychological conditioning. But it would seem that the only explanation as to why the majority feel disabled when in the presence of the Blue, that it is more due to trauma that is programmed into their DNA, verses actual personal experience.

Though there is proof that hereditary trauma does exist, there is also the possibility of Hereditary Enlightenment, as well. If a parent is living life constructively and rational, then naturally their offspring will inherit this way of thinking. Concerning this, it is important that whatever actions we chose to make in this life, that we do so constructively so that our children will inherit constructive thinking patterns from the beginning.

Other Unknown Environmental Factors:

Similar to the emotional state of the family, the environmental state of the birth may also play a role or leave an imprint on a baby when they are born. For example, a child born in the desert but relocated to a colder climate may find that they have a bizarre attraction to the desert, even if they have only been there when they were born.

There are many different ways that our environment affects how we are coded. But since we are exploring the world of Numerology, we will next learn how the Program Title or Name ties into Numerology and developing our identity.

There is a rule to deciphering a code. This rule is simply that the only way to decipher a code is when you have a key. Once you have the key, cracking the code becomes easy. The key in numerology is a chart that groups each letter under a specific spiritual number (spiritual numbers are the single-digit numbers). Zero is not included in the chart.

Alphabet – Number Decoding Key Map

1	2	3	4	5	6	7	8	9
A	B	C	D	E	F	G	H	I
J	K	L	M	N	O	P	Q	R
S	T	U	V	W	X	Y	Z	

Using the key map (above), you would replace every letter in a name, with its appropriate spiritual number. For example:

Susan = 1+3+1+1+5

Once you created the formula, the next step is to solve it.

1+3+1+1+5 = 11

The next rule is that all multiple digital numbers must be simplified into a singular spirit number:

11= 1+1 = 2

Now we have found the traditional spiritual number for the name, which is **2**. However, in the example that we went through, Susan would actually be 11 not 2 because 11 is a Master Number.

Master Numbers

A master number are numbers that vibrate the mind's eye and trigger power influxes in your spirit. There is a total of three master numbers, which are: **11**, **22** and **33**. The reason there are only three master numbers and not more, is because they represent the duplication of creation triangle, which is formed from 1, 2 and 3.

- **11** represents Intuition, Psychic and to be in tuned with Divine Energy.
- **22** represents the materialization of thought, success and the energy of a creator.
- **33** represents the creation or the product of 11 and 22.

In short, numbers are used to calculate or identify personalities and hidden codes. Here are some examples of Kemetic deity numerologies:

Deity Name	Formula	Personality description
Tehuti	2+5+8+3+2+9 = 29 = 2+9 =**11**	Intune with Divine Energy & the Cosmos
Ma'at	4+1+1+2= **8**	Infinite Energy & Balance
Ra	9+1 = 10 = **1**	Singularity, Self, Intent

These names have existed for thousands of years and have been put at risk of mistranslation. And yet, when we calculate the names of these deities, the numerology matches up to identify their characteristics, perfectly. As we have learned, Tehuti was the deity who was able to pull concepts out of the Divine Energy and to bring them forth into materialization. Thus, he would have to be Intune with the Divine to do so. Ma'at is the deity that represents infinite justice and balance, as her numerology reading clearly states. And Ra, the deity of the Sun who is of masculine energy and spirit and intent, has a reading that represents the same thing.

Take a moment to write out your name, formula and decode your personality to see if it matches:

Name	Formula	Description

Other popular numbers, nicknamed Angelic Numbers, are:

- **1:11** means it is time for a spiritual awakening.
- **2:22** means that you should have faith and trust.
- **3:33** means that your ancestors and guardians are nearby and are ready to assist.

- **4:44** means that you should be encouraged and reminded that you are loved.
- **8:08** is a combination of numbers, in which 8 represents infinity and balance, while 0 represents Divine energy. This number is typically grouped with finances and the materialization of something. The manifestation of abundance, while being in line with the Divine. This could also mean that you are an inspiration to other people.
- **11:11** means that you are on the road to success. You are in tune with the Divine Energy.
- **12:12** means that you are on the verge of spiritual awakening.

In short, numerology plays a huge role in identifying the genetic code that makes up our personalities. Additionally, these numbers are used throughout the metaphysical culture and in the normal lifestyles we live. Sometimes these numbers are messages to enlighten us and sometimes these messages are warnings. The world of numerology is fairly complex. It dates back before the bible was created and has many more codes than what we can imagine.

History of Mathematics

The earliest recorded form of mathematics is documented to have started in 35,000 BCE on the continent of Africa. More and more archeologists have discovered that there were advanced civilizations that predated the carbon dating of 35,000 BCE. The oldest skeletons in the world were found in Northeast Africa and along the Rift, central east Africa and were carbon dated as being over 7 million years old! Minus pseudoscience attempts to reform these images to appear Ape-like to support the illogical conclusion that we stem from Apes; these fragmented skeletons were proof that early humans existed well beyond the original dates predicted by science. In all cases where these ancient skeletons were recreated for global presentation, the artists reconstructed these artificial representations of prehistoric bodies, of a handful of fragmented bones and shards. From a small number of shards, thee artists molded images of Ape-like humans with dark brown skin! The final representation of a black Ape-like woman is an example in which the push to support the false notion we were once monkeys is racist, as there was not enough evidence to support neither the complete structure of the bone (no matter how fancy the designer's title was) nor the coloring.

It is important to note that in this theory that humans came from Apes, Science has failed to provide an example where Nature repeated itself, if at all. According to pseudo-science, humans came from Apes and evolved from black to white. However, there are a few contradictions to this theory, in which, Apes have fur. African people have kinky hair. It is illogical for nature to evolve a species to change from straight fur-like hair to kinky than back to straight fur-like hair. Additionally, Apes have thin lips. Just like the hair example, it is illogical for Nature to evolve a species from one thing to another than back again. Also, when Apes are shaven, they are not black (or dark brown), many are pinkish gray. Furthermore, when Nature evolves a species, it is making the previous version redundant. The older version dies off, which is NOT the case with humans and Apes. Additionally, the cognitive wiring of an Ape brain is drastically different than the human brain. The left and right brain functioning are backward and different from that of a human. This is also illogical for Natural evolution, as regardless of the species move to adapt, the basic brain process remains the same. Finally, there is a complete collection of Homo Sapien evolution that was found in the Rift of East Africa. It showed the progress of earlier Homo Sapiens evolving into modern ones, where the forehead broadens as the brain broadened.

Darwin's theory of evolution was based on microevolution, not macroevolution! Microevolution is where a species will adapt on a smaller scale to adjust to the new environment. Darwin's theory was based on Finches and how one Finch was yellow in one region of the world, while the other one was brown. Darwin concluded that when the finches migrated over the Atlantic, they adapted to their environment by changing their color. Macroevolution, on the other hand, is when a species evolves into another species (which still fails to be proven). This is what the concept of the Dinosaurs and Humans evolving from Apes is based on. We can find examples of microevolution in human history (if we refer back to the study of generations). In chapter Generational Challenges, we learned that the mental processing of people differed per generation due to upbringing and their environment. This is an example of microevolution. Another example of microevolution is that when black people are constantly deficient in vitamin D, their skin tone lightens, and their next-generation becomes lighter than they would if they were born in a sunny region. An example of that can be found in the multiracial children born in the Northwest, where people are all deprived of natural vitamin D. The multiracial children of this region are drastically lighter in complexion (in many cases, almost appearing white). When compared to multiracial children in regions of enriched sunlight, the children are darker in complexion. This is an example of microevolution.

Outside Scientific Racism, other historical artifacts contradict the theory that we came from Apes, for example, there are million-year-old electric generators made from stone or the 10-million-year-old hammer or the million-year-old sculptures that were used as ancient compasses or makeshift skin boats that were created millions of years ago for ancient humans migrating out of Africa to south Europe. All of these artifacts suggest that humans have been sophisticated creatures from the dawning of time, which separates them from Apes. Included in the list of prehistoric "phenomena" is the use of mathematics and early commerce activities, which is particularly found in the 40,000+-year-old iron mines and melt smelting trades in South, Southeast, and Southwest Africa.

As we discovered earlier, mathematics can be found in everything creative. Therefore, we can assume that mathematics has been practiced quite regularly throughout human history before an abstracted formula of the concept was created. Deep within the mountain range called Lebombo, in Swaziland, archeologists discovered that ancient Khoi Khoi tribes used a baboon fibula as a means to calculate measurements. It is difficult to figure out exactly what this prehistoric man-made device was used for. Some speculate it was used to monitor the lunar or menstrual cycles, while others speculate that it could have been used as a measuring stick. Archeologists named the counting device **Lebombo Bone** since it was found in the Lebombo Mountain range.

Fast forward to 25,000 BCE in prehistoric Europe, the construction of the Venus Figurines was calculated to have started. The "<u>Venus</u>" figurines are miniature statues of a full-figured woman;

with wide hips, large buttocks and large breasts. The shapeliness of the feminine image matches more closely to the Khoi Khoi women, particularly Hottentots (like the famous Sarah Baartman, even down to the dense hair coils), instead of the type Caucasian European shape. As mentioned earlier, these miniature statues were also used as a prehistoric compass as well as for idol worship. The lines carved into these figures appeared to point north, south as well as precise degrees. These lines appear both in the front as well as in the back of the sculpture. According to researchers, the lineage markings are identical to the Ishango and Lebombo Bones. The Venus sculptures were found throughout Europe and Asia and marked the prehistoric human migration pathways. Since the markings on these figurines are very similar to the Lebombo Bones, it can be assumed that Europe, Asia, and Africa shared the same astronomical theories and methods to calculate mathematics, as well as longitude and latitude, roughly 25,000 years ago.

The Ishango stick is calculated to have started in 20,000 BCE. This stick is similarly crafted like the Lebombo Bones, however, it demonstrates more advanced forms of mathematics. The Ishango, like the Lebombo, was constructed on a Baboon's Fibula. The Ishango stick was discovered in the modern-day Democratic Republic of the Congo. The location of this discovery is very important because the DR of the Congo's eastern border covers part of the East Rift, which is speculated as being the starting point of the homo sapiens as well as one of the states of the Kongo Empire.

Unlike the Lebombo Stick, the Ishango stick has clusters of markings, which represents a variation in quantities. One side of the stick has only prime numbers from 10 to 20, while the opposite side contains the sum numbers, from numbers that are added and subtracted by the prime.

Roughly around 7,000 BCE, the early Kemetic Tribes would use clay tokens as counting devices, a technique that was simultaneously being used in Ancient Asia.

The Kemetic calendar is first recorded to have started roughly around 5,000 BCE. The Kemetic civilization, started long before, roughly 10,500 BCE, however, the recorded calendar is said to have been created in 5,000 BCE. This is also parallel to the first Kemetic stone hieroglyphs. The Kemetic calendar was 365-days long, with 3 seasons that were stretched out for 120 days.

In 700 BCE, in a place called Yeha (Ethiopia) counting games were practiced by natives. The most common game is Mancala. It is suspected that these mathematics games were practiced many years before, since the dawning of humanity in Central Africa. The game is played using a wooden board, with 12 pockets carved; 6 holes in 2 rows. Each hole starts with 4 stones or marbles. Traditionally, there are no pits at either end of the board. The first play starts by picking up the pieces in one of the six holes on that player's side of the board. The player moves counterclockwise, placing one stone into each hole the player passes. If the player reaches their storage drop, the player places a stone in it but skips their opponents. Once all the six holes on one side of Mancala board is empty, the game is over. The player who has the most stones wins.

Meanwhile, during the so-called dark ages, when the Moors ruled Europe, the Moors brought with them the concept of Algebra, which stemmed from Northern Africa during the early ages of the Common Era (still before the construction of the bible, Qur'an or Judaism texts). This further supports the fact that mathematics remained an intricate part of African culture but was present within the context of the African environment.

To conclude, Africans were not idiots or unsophisticated. Additionally, we can use mathematics to prove that the "human from Apes" evolution theory is false. Not only is there a collection of humanoid skeletons that evolved, found in Africa that clearly distinguishes the difference between humans and Apes. But also, these prehistoric humans were capable of calculating early forms of geometry (as we see in their artifacts), arithmetic, statistics, economics, and algebra. Math has played an enormous role in human development since before the symbols representing the values of numbers were created. We used mathematics to guide us as we migrated around the world. We used mathematics as a form of entertainment, and to help us develop our homes. The ancient African villages, towns and city structures, demonstrate how much math played a huge role when calculating the layout of their homes, calculating menstrual cycles, creating basic calendars and managing a trade.

And even beyond this, prehistoric Africans used mathematics to calculate and map the stars. As we have learned that even the most modest tribe, still demonstrated an understanding of calculating a navigation tool using mathematical metrics. These accomplishments are not the accomplishments of an idiotic species that is depicted throughout western science as a means to push scientific racism. These are the accomplishments of a very intelligent species that was able to develop a complex civilization before they could construct complex technologies. Concerning spirituality, the proof that human advancement was millions of years old instead of thousands, supports the explanation as to how humans were able to develop the skills to construct so many creative works. As far as "leaps" of human advancement goes; there is a pattern where discoveries that are associated with specific demographics (races) of people are dismissed in the history books. It is more beneficial to a racist system to promote the concept that "magically" humans evolved and that advanced civilizations (that can not be explained away as being white) are presented as "vanishing." The reality contradicts these racially motivated lies, in which these "vanishing" civilizations were African nomadic tribes and these technological advancements were found in Kemet (particularly after the Greeks invaded Kemet). Artifacts and history show the slow mental evolution and development of mankind. Additionally, technical advancement shifted in regions where southern prehistoric tribes (homo sapiens) and northern/eastern tribes mingled. Shortly after these tribes came in contact with homo sapiens, these northern/eastern tribes suddenly began conducting in agricultural practices (particularly in the Great Mixing of Human species, in the Mesopotamian region roughly 40,000 years ago). For other species of humans, this was a sudden leap in technical advancement, but for the southern tribes, this technical and critical thinking abilities were hereditary. The continent of Africa was a ripe place for humans to form. There was a vast amount of resources to explore with, from harvesting salt from the ocean to working metal ore. The first iron mining and melt work were started some 25,000 years ago (if not more). Africa, at this time, was also rich in animal life where the homos sapiens were able to study and learn from them. What these prehistoric tribes learned; they wrote on Traveler Boulders (which are ancient boulders with prehistoric carvings – pictures with instructions about the area). So, when other roaming tribes would approach, that they were aware of both the dangers and the history of the land. These tribes did not dwell permanently in caves. They roamed the land using the stars as their guide. Millions of years later, their footprints are permanently etched into the ground where they walked, in Africa.

In this setting, it is only logical that these species of early humans would advance.

Star Gazers

AS · TROL · OGY

(noun)

The study of the movements and relative positions of celestial bodies interpreted as having an influence on human affairs and the natural world.

Our ancestors taught us that the energies of the Universe influence how we behave, think and develop. This concept isn't too farfetched to believe, as we see examples of how celestial bodies influence our world daily. For example, the moon influences how our gravity works and triggers the planet's wind and water currents. The Moon even controls the animal and human reproduction systems. On an even broader scale, the other planets in our solar system influence our planet and protect our planet from foreign celestial bodies (asteroids and debris) that may randomly fly into our solar system. Planets like Jupiter, Saturn, and poor Pluto, take the bulk of the hits from random debris. Their gravitational pull brings in most of the objects that may float in this direction. It is almost as if their orbit route and timing, was timed perfectly to ensure life would grow on Earth. Or perhaps it was coincidental how the timing of the movements, as well as the shape of these celestial bodies, were perfect to ensure that life could grow on Earth. Whichever way you look at it, it is amazing.

The study of the stars was an essential part of our advancement. Our ancestors studied the stars so much that they were able to map the sky, without high-tech telescopes. It is hard for many of us to imagine how they could have such an accurate depiction of space to the point where we are still discovering things that our ancestors were already aware of, like Sirius. The irony is that we are supposed to be more advanced, and yet, we are just rediscovering what our supposed *primitive* ancients already knew. Our ancestors were the first to fully understand the concept of *As Above, So Below.* They used the knowledge that they gathered from studying the Universe to map out life down to the smallest molecule.

"The ancient Egyptians have developed a very complex religious system, called the Mysteries, which was also the first system of salvation. As such, it regarded the human body as a prison house of the soul, which could be liberated from its bodily impediments, through the disciplines of the Arts and Sciences, and advanced from the level of a mortal to that of a God "– George G. M James (*Stolen Legacy*)

Our ancestors were excellent, indeed. They knew that what they saw above them would be reflected in their reality and for things they were unable to see.

To ensure that their discoveries would remain an inspiration throughout time and space, the Kemetic people created a celestial map that also told the stories of these celestial beings, who became their deities. This celestial map has been used by other religious systems, including Christianity. This is important to take note, because Astrology (and Astronomy) are also a part of the *666 Abrahamic Sins*, as it states:

"Then He said to me, *"have you seen this, O son of man? Turn again, you will see greater abominations than these."* So He brought me into the inner court of the Lord's house; and there, at the door of the temple of the Lord, between the porch and the altar, *were* about twenty-five men with their backs toward the temple of the Lord and their faces toward the east, and they were worshiping the sun toward the east." *Ezekiel 8:16 (New King James Version – NKJV)*

"Whoever seeks knowledge from the stars is seeking one of the branches of witchcraft..." (Abu Dawood; Saheeh; *Qur'an*)

The Ezekiel scripture is a direct reference to Kemetic worship of Ra, in which the worship of the Sun was an abomination. This is ironic because the cross (which is the symbol of Christianity) is one of the most popular Ancient symbolic references to the Sun in the center of the solar system (where the two lines meet). Before labeling this as coincidental, it's important to note that historically the cross (and all of its many different versions) were always a reference to the Sun. The Christian Sun worshipping can be dated back to the god **Shamash** (which is hypocritical as Babylon was condemned by Abrahamic scriptures). There are even other studies that source the Christian Sun worshipping back to Kemet! In early Christian churches, for example, there is the Ankh symbol on the wall and the boat that carries Ra throughout the night to rise in the morning. The symbol of the cross can also be found on the walls of Kemet, this includes the Swastikas cross, which is a depiction of the four Elder Gods in the zodiac surrounding the Sun. In more modern versions of the cross, where Jesus is shown crucified, the Halo is the representation of the Sun's halo. We can see this by looking at older versions of the cross with Jesus crucified on it, where his head is surrounded by the rays of the Sun, depicting his heavenly power and his symbolic reference to the Sun. If that is still not enough proof (to all those skeptics out there), we can go even further back to cross artifacts that have the Sun carved into the exact position where Jesus' head will be strategically placed in later years. As the Jehovah's Witnesses like to point out, in Jerusalem, it would be impractical to waste wood building giant wooden crosses to execute prisoners. Wood was a limited commodity; therefore, it would be impractical and very unlikely. In light of this fact, the symbol of the cross was implemented into Christianity by another means, as was the symbol of the fish (the symbol of the fish can be referenced back the Assyrian deity named **Dagon**).

The Assyrians and the Babylonians were heavily condemned by Abrahamic scriptures, and yet, ended up becoming the main symbols representing these monarch organizations. This also includes the Muslim symbol of the star and a slighted moon, which comes from the Sumerian Goddess named **Sin** (who was the Goddess of the moon whose symbol was the Moon and a single Star). The various nations that resided in Mesopotamia that were condemned by the Abrahamic deity (Assyrians, Babylonians, Hittites, and Sumerians), were heavily incorporated into the religious doctrines, culture, and ideology associated with Abrahamic cosmology, including astrology. Even the Jewish Star of David can be traced back to the Babylonian Sun God Shamash. Meanwhile, the Catholic priests wear fish hats that are very similar to the hat of Dagon (the Assyrian Fish God) or the hat of Kemetic priests.

If all factions of Abrahamic cosmology, a reference to symbols sourced from nations condemned by Abrahamic deities, could we still consider these connections to be a mere coincidence?

No. With so many similarities and clear connections to ancient cosmologies, we cannot dismiss these connections as mere coincidences. Instead, we have to accept the reality that all Abrahamic factions are Sun, Pagan and Astrology Worshippers!

Oh, but the Catholic church was a massive marketing organization that would adopt various religious elements for locals, to convert them to Christianity. I hear the skeptics say.

There is a pattern in Abrahamic scriptures where things are openly condemned as Sins but are later supported in the actions of the Abrahamic heroes, making this a pattern of contradictions. For example, it is Sinful for a woman to be a prostitute, but regardless of this Sin, the male prophets went to prostitutes or collected concubines; not wives.

Abrahamic Cosmology is no different than any other cosmology that is based ideology is based on cosmic activity that is blanketed with political agendas. This is especially true for Abrahamic factions since the stories in Abrahamic cosmology were plagiarized off other cosmologies. It is possible that during that Council of Nicaea (the council that created or put together the Bible) were unaware that the stories that they copied and pasted were about the cosmic activity. But regardless of this possible ignorance, what they didn't know then, they openly know now, as their religious structures depict the Zodiac and Solar Activity. This solar depiction can be seen on the Vatican floor, where the seat of the powerful people is situated in the middle, where the Sun would be. This ideology of power being in the center of a circle is eerily echoed by the white supremacist Black Sun and the Swastika symbols. But undoubtedly represents the same concept of the cross.

Conceptualized of Astrology

We may not be able to see the stars on a regularly. But we can create our own "***celestial star map***" by using our tarot cards to help us understand and see what our ancestors were able to see. Using the tarot deck to map out the star map, will help build a deeper understanding of the card characteristics as well as understanding the characteristics of each decan.

The Kemetic people not only outlined a celestial map, but they created an accurate calendar system and clock system that we still use today. These calendars and clocks were painted and carved onto their buildings and even on their coffins. The clock system, ironically, consisted of 12 houses and 36 Decans of 10 degrees. Marshall Clagett, in the book called *Ancient Egyptian Science: Calendars, Clocks, and Astronomy*; stated that:

"… the principal astronomical diagrams before the development of the standard celestial diagram were those of the diagonal clocks of the coffins… in the case of the clocks dependent on risings… the ideal clock… in its columnar structure included the names of 36 Decans applicable to the 36 decades of the year plus 12 Decans for the epagomenal days." – page 108, para. 3

This practice of the use of multiple meanings for each symbol was a common practice amongst the Kemetic people. In this case, the Decans not only represented a *Present* moment in time but also represented decades of the year. Clagett continued to state that:

"In the ordinary celestial diagram which we are now considering all 36 Decans of the diagonal clocks (for more or less than the 36) were listed but not in the diagonal form used on the coffins. Usually, one (or occasionally more than one) star accompanies the decanal name as a determinative. If more than one, the stars indicated an attempt to represent a decan consisting of more than one star. Furthermore... groups of stars appear in the columns below the Decans. However, such star groupings or clusters are of little assistance in identifying the constellations they represent because the number and arrangements of stars for each group vary widely from monument to monument. Exceptions are those below the Decans related to Orion [Osiris/Ausar] and Sirius, which can be identified, at least roughly. The names of the deities associated with the Decans are usually included, but again not with great uniformity. Sometimes depictions (and not just the stars from which depictions were conceived) of the gods or possibly larger constellations associated with a number of the Decans were also included, e.g a bark, a sheep, an egg. Orion [Osiris/Ausar] in a bark, and Isis-Sothis in a bark." (Clagett, M., *Ancient Egyptian Science: Calendars, Clocks, and Astronomy*; page 109 para. 2)

A few things that we can learn from Clagett's statement is that the depiction of the cluster of stars varied, except for the stars that represented Ausar and Sirius. Additionally, we learn that these constellations were depicted with everyday objects and creatures that the Kemetic people were familiar with. Many of these depictions are depictions that we still used today, within modern Zodiac symbols. It is important to take note, that the names of the Kemetic deities were changed into Greek names. For example, the term Orion is Greek. The original name for that deity is Ausar or ASR (as it was written on the walls of Kemet). While the original name for Sirius is Sopdet/Sepdet (which means "she who is sharp"). Sopdet is a Kemetic deity. She is said to be the cradle of human knowledge and was regarded as the most important star in Kemetic astrology.

"Over twenty times brighter than our sun and twice as massive, its brilliant white color are tinged with blue and purple. All the colors of the rainbow sparkle from Sopdet (Sirius) when observed low on the horizon during certain atmospheric conditions. Some mysteries regard Sopdet as the true light and source of all life including our sun – "shadow" of the great star – which illuminates the illusory physical world; whereas the great star, Sopdet, keeps the true spiritual world alive. Sopdet has crossed from the east bank of the Milky Way where it resided some 100,000 years ago to the west bank of the celestial Nile River where it currently rests." – Mutere, M., *Auset's Star – Humanity's Mystery*

The great pyramids of Giza are built along Ausar's hunting belt, these are most commonly recognized as the three stars on Orion's belt. In Kemet, these stars were known as the "three kings" or the "three wise men/magi." The symbol of these three kings can be found visually as well as written in popular stories, such as the story of the birth of Jesus, which is outlined in Abrahamic scriptures. The story of these Wise Magi reads:

After Jesus was born in Bethlehem in Judea, during the time of King Herod, **Magi from the east came to Jerusalem** and asked, "Where is the one who has been born King of the Jews? We saw his star in the east and have come to worship him."

When King Herod heard this he was disturbed, and all Jerusalem with him. When he had called together all the people's chief priests and teachers of the law, he asked them where the Christ was to be born. "In Bethlehem in Judea," they replied, "for this is what the prophet has written:

'But you, Bethlehem, in the land of Judah,

are by no means least among the rulers of Judah;

for out of you will come a ruler

who will be the shepherd of my people Israel'."

Then Herod called the Magi secretly and found out from them **the exact time the star had appeared**. He sent them to Bethlehem and said, **"Go and make a careful search for the child** *[son]. As soon as you have found him, report to me, so that I too may go and worship him."*

After they had heard the king, **they went on their way, and the star they had seen in the east went ahead of them until it stopped over the place where the child [son] was**. When they saw the star, they were overjoyed. On coming to the house, they saw the child [son] with his mother Mary, and they bowed down and worshiped Him. Then they opened their treasures and presented Him with gifts of gold and incense and myrrh. And having been warned in a dream not to go back to Herod, **they returned to their country by another route.**

(Matthew 2:1-12, *New English Translation, 2005*)

This story of the three Magi are rhetorics that tell the story of the celestial movement, or in other words, they are verbal star maps. These verbal star maps are found in the stories of the Orisha deities and the Zulu deities and the ancient Indian and Asian deities. Since Abrahamic Cosmology is plagiarized from these sources, we can only assume that this is also found within modern religious doctrines.

The Kemetic people told the story of Ausar (the constellation) which is physically located along the equator. As the story goes, Anubis assists Auset as she searched for the pieces of her husband, which has been "scattered" by Set (Ausar's brother). Once Auset puts her husband Ausar back together, he symbolizes rebirth or

Spring (which correlates with the "birth of Jesus" as he was born in the Springtime when the animals were giving birth).

The three stars on Ausar's belt, are the three Magi's who prophesize Sopdet or the "*Morning Star.*" When the Morning Star shines, her Sun (son) will soon rise. The Sun, Ra, is the Sun (son) of Sopdet (the Morning Star), which follows his mother into the morning by rising in the East. Once the Sun (son) has risen, the three Magi's (the three stars on Ausar's belt) return to their home by going a different direction than the one they came.

In various social conditions, astronomical activities play a huge role in our civilization. Including how we calculate time and our belief system. These celestial bodies also play an important role in human nature. The depictions of ancient deities as animals and objects, as stated earlier, have been revamped as the symbols we most commonly associate with zodiac and planets. Our understanding of how pictures slowly convert into symbols and letters (which is outlined in the Chapter: **Symbolism**) can help us to understand how the original symbols of these deities became what we are most commonly familiar with.

Regardless of the changes made to these symbols, the characteristics of the Elder Gods, the Houses, and Decans, remains the same. Astronomy is embedded throughout modern religion. And the symbols created to represent these cosmologies represent ancient celestial deities (primarily the Sun, Moon, and Star Constellations).

Another example, of Astronomy in the biblical text, would be the mention of the *"four corners of the world"*; which is a reference to the four Elder Gods or the four elements:

Revelation 7:1 [1] After this I saw four angels standing at the ***four corners of the earth***, holding back the four winds of the earth to prevent any wind from blowing on the land or the sea or any tree.

Ezekiel 7:2 [2] "Son of man, this is what the Sovereign LORD says to the land of Israel: "'The end! The end has come upon ***the four corners of the land***!

Isaiah 11:12 [12] He will raise a banner for the nations and gather the exiles of Israel; he will ***assemble the scattered people*** of Judah from the ***four quarters of the earth***.

(from the New International Version, NIV, 2017)

In Science, we learned that there is no corner of the Earth, as it is round. However, according to the celestial maps, the celestial "world" was divided up into four quarters, each representing the different equinox and elements. Therefore, when we read the line about "I saw four angels standing at the four corners of the earth"; it begins to make more sense. The Swastika, for example, represents the four corners of the zodiac map; which is divided up into four parts to represent the four realms of the Elder Gods with the Sun in the middle.

Another key astrology connection found in Abrahamic cosmology is that there are twelve houses are also embedded in this religious story. The twelve zodiac houses are represented by the twelve disciples/apostles. There are about 31 references to the twelve disciples in these scriptures, so, I will not analyze each one of them. The scripture I would like us to analyze is Revelation 21:14, which states:

And the wall of the city had twelve foundation stones, and on them were the twelve names of the twelve apostles of the lamb.

(from the King James Version, 1611)

The first key point that we should pay attention to, is the mention of a city that had twelve foundation stones. In most cases, cities and towns were built circular, it was a very common practice in our human story. The fact that there are twelve foundations around this city mysteriously mimics the structure of Kemetic clocks and celestial maps that Clagett pointed out in *Ancient Egyptian Science: Calendars, Clocks, and Astronomy*. The next key point about this scripture in Revelation, is that "the twelve names of the twelve apostles" were written on these twelve foundations, which also is the exact practice that the Kemetic people did (literally, this is what Kemetic people did when they built their temples), which was to write the name of their deities on the base of these foundations (Houses). Perhaps a future volume with uncovering the excellence of blackness will be to go through an entire book that is dedicated to decoding the bible/qu'ran/Tanakh and Talmud, as there is so much that can be uncovered about the celestial indirect biblical references. At this time, though, the other key point that we will analyze about the above scripture is the use of the term Lamb.

A lamb is a biblical reference to sacrifice and is commonly used to refer to Christ. As we have learned earlier, Christ is the Sun. Therefore, the twelve Apostles of the Lamb could only be referencing to the twelve Zodiacs of the Sun.

In short, we should weigh the actions of an organization above their words. In one stroke, Abrahamic scriptures condemned you from looking up at the sky. While with another stroke, the entire Abrahamic ideology was formulated from astrology. Our ancestors were able to advance beyond our understanding because they studied the stars and understood the concept: As Above, So Below. Therefore, so that we may advance, we must go out of our way and work even harder to study the stars as our ancestors did. By studying the stars, we will be able to reconnect to cosmic order and begin to understand the depths of our awareness.

Try it! Take a weekend to go camping. Take with you a telescope and a sketchbook. Under a clear night sky, try to map out the constellations that you are able to see. Then when you return back to metropolitan life, research the backstory of those constellations. More than likely, the indigenous people in fused the history of their people with the story of the stars.

The True Meaning of the Cross

A Cross is created with two lines, in which one is vertical, and one is horizontal. The intersection where the lines overlap is the center point of the symbol. Sometimes the lines are of equal length and other times the vertical line is slightly longer. Regardless of this subtle difference, the symbol is the same (if anything, the region where the longer line goes out further than the rest, can also represent a simplified arrow, pinpointing a specific direction, like the minute hand on a clock face) What we no longer see in this symbol, which existed in the past, is the circle that the cross resides in. That hidden circle represents the celestial map or a star map.

The lines of the cross act as dividers of the star map, which divide the circle up into four parts. Each part represents an Elder Gods or the elements: **Fire, Water, Air,** and **Earth.** These regions of the Elder Gods are each an Equinox. Each Equinox hosts "three houses" and within those houses, lies three Decans of 10 degrees.

Fig.

- **Equinox = Elements or Elder Gods (Fire, Water, Air and Earth)**
- **Houses = Zodiac Signs**
- **Decans = The various characteristics or stages of characteristics, within the house**

In the center of the circle, is the Sun. Just like a Sun Dial, the minute hand moves clockwise along the face. As the minute hand moves, it goes in and out of the different Houses, Decans, and Equinoxes.

With much of a strong connection to Solar Activity, it makes you wonder:

Why condemn astrology if the premise of cosmology is astrology?

The simplest answer to that question would be: if society advanced to god-like levels from studying the stars, then it is only natural that an oppressive system would no longer want you to lookup. Most of us live under heavy light pollution, and never see the stars, let alone study them.

Kemetic Star Map

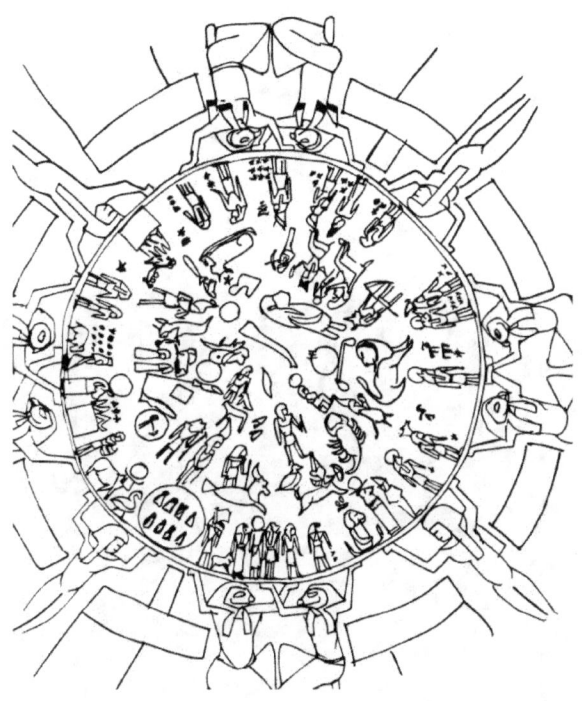

Knowing vs Believing

In the first section of this book, we have explored the concept of excellence. We also explored the dangers of living in extremities, particularly the unnatural removal of "bad" to live purely "good."

Abrahamic doctrines teach its followers to follow the faith and to turn away from science. In moderation, there is some truth to this, as science also has extremities that lead us to become delusional about reality. But with that said, there is value in science, and it is a necessary practice that keeps us able to develop our minds and to remain rational. The Abrahamic extremity is to isolate yourself from critical thinking and self-development. This is a similar ideology that we find in the prison system and military systems that we explored before, where the system is designed to keep prisoners and military personnel mentally stagnant in the regions of the brain that formulates individuality and self-growth. In Abrahamic religions, the main theme is to prevent the follower from thinking for themselves. Ultimately, the perfect follower of these systems would not utilize critical thought or have a desire to be an individual at all. In other words, to never question anything, just obey.

The problem with this concept if of absolute submission is that every time the religious doctrine is "updated" new political agendas are added. The King James Version, for example, has been repeatedly updated, and in each of those updates the scriptures are reworded and the name of God is changed to lord, making the god in the heavens and the Abrahamic Christ indistinguishable from each other, to further support the ideology of the Trinity (which is never mentioned or referred to in biblical texts – as of yet). This conversation of the Abrahamic deity's name to LORD and the Abrahamic Christ's name to Lord, has been a major push since the 1970s, but was originally started when the book was created for King James I and IV (King of England and Scotland), which is outlined in the prefix of a KJV. But regardless of this information, most followers will not read either the prefix or the history of the process of how these various versions and translations became published.

I am an inquisitive person and so, I began to research and ask about the background to how these religious books were created. I was shocked by some of the responses that I received. What I found even more curious, was that a lot of the responders shared or echoed each other's answers, as though they were reading from a script. I found a similar universal response when researching the signs and confessions of domestic abusers, sociopaths, racists, extremists, victims of PTSD and other mental syndromes. This Standard Response behavior seems to be a sign when a person is suffering from some form of mental suppression. In this case, some of the Standard Responses were:

"I don't read any source that does not come from God."
"The history of how the bible was created doesn't matter."
"The Word is the Truth, that is all that I need to know."

"You should never question God."
"It doesn't matter."
"If the information doesn't come out of the bible, then it is worldly and corrupt."
"Why would I research the history of the bible?"
"I know where the Qur'an came from, it came from god. That's it."
"If God wants me to read something else, he will let me know."
"The history of the Qur'an is written in the Qur'an. Read it and you will know the history of how it was made."
"The New World Translation is the truth and the most accurate translation of the bible. It is not important to focus on how this translation was created. What is important is that it was created and will bring you closer to Jehovah."
"I don't read anything else but in the bible."
"You will go to hell for questioning god."

According to history, Jesus would have died at 37, not at 33, because while the bible was being constructed by *the Council of Nicaea,* in 325 CE. During this time, the Council forgot to calculate a 4-year adjustment found in the Hebrew calendar, which in turn through the timeline of the story. The Qur'an, on the other hand, was constructed in 610 CE by Abu Bakr, which was based off the Arabic timeline (the Qur'an was later translated by an English man named George Sale in 1734; this version became the most common version of the Qur'an), making the timeline more accurate than that of the bible. When researching information about religion, you will undoubtedly come across articles and books that are filled with dogma.

Dog·ma (noun)

Dogma are principles that have been put in place by an authority that is treated as incontrovertibly true. It is a mechanism used to push or formulate the ideology or belief system of an authority system, and it cannot be changed, questioned or removed without affecting the foundations or model of the authority system.

To be successful as a critical thinker, it is important to understand that there is a difference between a Belief and Knowledge so that you can learn how to identify what is dogmatic and what is not.

For example:
Belief – Man was born from a Man.
Knowledge – Man was born from a Woman.

The difference in these examples is that with the knowledge we have proof, while with a belief we do not have proof. To put it differently for the suborn folk... We all have proof that our (men and women) birth was through a woman. The Proof can be traced back through documents, video, numerous unrelated witnesses, DNA, scientific data analysis and common agreement that this was an actual occurrence.

On the other hand, in the religious text, we do not have any proof except for theories. Additionally, these religious texts reference to themselves and not externally. In educational practices, a document is not considered credible unless it is sourced with external citations and references to support the objective. Religious doctrines do not have external citations or references, and references to external "sources" are vague.

One example of vagueness can be found in the story of Moses, where: Moses was in Egypt and lived in that area under the reign of a pharaoh.

Do you know how many pharaohs ruled Egypt? 170!

170 pharaohs ruled Egypt. Strangely, the scripture wasn't more specific. In light of this, the region is known in the bible as "Egypt" wasn't named Egypt until the 1800s of this Era. While Ethiopia was named Ethiopia during the British Colonization of Africa during this Era. Before that, the region we know as Ethiopia was called Kush, and Egypt before colonization were called Kemet (KMT). Both were renamed by their invaders, and in the religious doctrines, these regions are taught to have "always been" named their Grecian names. The push to convince religious followers that Ethiopia and Egypt are the true names of those regions is a form of dogma.

Another example of vagueness is found in the story of Jesus, where: he was crucified by the Romans under Caesar's rule. Roman ruled over Jerusalem for over 380 years (from 63 BCE – 313 CE; there is a gap of a few years when the Eras were shifted)! Crucifixion was a common act of execution by the Roman authoritarians, which means hundreds of prisoners were executed in this fashion (which was not done using excess wood to form a cross, due to a shortage of wood in the region). There were numerous Caesars in that period (as we have had numerous presidents during the existence of the United States), and out of those many Caesars, there are about 12 that are up for constant debate.

Additionally, another example of vagueness is found in the story of the Great Flood around 8,000 BCE. There is no proof that the Great Flood ever existed. Yes, I hear the naysayers claim that the Ice Age could count as a Great Flood (as Ice is water that's changed its form). However, if that is the case, then it may be important to point out that the most recent Ice Age was about 22,000 BCE and it only covered 1/5 of the planet... not the whole world.

Though these are truths that denounced the authenticity of Abrahamic dogma, I also encourage you to research for yourself, as what is true today can be disproved tomorrow.

"What is right today, may be proven wrong tomorrow. That's why today, I am open to any changes that may happen tomorrow." – Dr. Yosef Ben Jochannan (from the video lecture "Belief vs Knowledge")

The truth is that there is no truth. Or as it is better said from the movie *The Matrix* "The truth is that there is no spoon."

From all that you learned in the first section of this book; you have come to understand that there are infinite perspectives about similar situations. Even the experience of reading or listening to this book, will not be received the same as the next person.

This is what makes sourcing from the Eternal Self so powerful! Ultimately, only you can conclude what is true to you. You are the only one who can conclude as to where your boundaries lie, if there is one, at all. You have been constructed to be able to decipher what is true for you from fiction, by using rationality and comparative analysis. This act of comparative analysis of daily experiences and information is called critical thinking. Critical Thinking is labeled as a deadly sin (or evil) by Abrahamic authorities. If we are denied the ability to think for ourselves, then we exist only to be puppets. If we only believe that life must have a purpose that is gifted to us by an alien, then we are more susceptible to self-inflicting because we have denied ourselves the right to see value in loving ourselves.

Returning to the bizarre side effects of mental suppression, you will also notice that it is very common for the oppressor to warp how things are perceived by those being oppressed. Oppressors tend to follow the same kind of oppressive behavior, regardless if they are an organization or an individual. When reviewing the 666 Abrahamic Sins, I found that the sins were eerily in support to the oppressive behavior of a domestic abuser. These abusive signs of a domestic abuser, according to the *National Coalition Against Domestic Violence* (NCADV) are:

- Extreme jealousy
- Possessiveness
- Unpredictability
- A bad temper
- Cruelty to animals
- Verbal abuse
- Extremely controlling behavior
- Antiquated beliefs about roles of women and men in relationships
- Forced sex or disregard of their partner's unwillingness to have sex
- Sabotage of birth control methods or refusal to honor agreed upon methods
- Blaming the victim for anything bad that happens
- Sabotage or obstruction of the victim's ability to work or attend school
- Controls all the finances
- Abuse of other family members, children or pets
- Accusations of the victim flirting with others or having an affair
- Control of what the victim wears and how they act
- Demeaning the victim either privately or publicly
- Embarrassment or humiliation of the victim in front of others
- Harassment of the victim at work

The only difference between an individual being an oppressor and an organization is the slight change in verbiage. Whereas an individual oppressor would tell their victim that "you are stupid." An organization would say "The Lord calls them stupid" or "God names them idiots" or "those that disobey are fools" and so on. The shift is found in the style of verbiage changing from the first person to third.

I do not doubt that these major religions play a huge role in the toxic masculinity found throughout the world. After all, if an individual has a Standard Response that stated that their religious book is the only thing they ever read or study. The result would be the development of a large population of people whose ideology is oppressive. If the only source of information that you allow yourself to be exposed to, justifies oppression, then naturally if you are a "true believer" you will no doubt become an oppressor.

> *"You imprison yourself with these nonsensical things."* Dr. Yosef Ben Jochannan (Belief vs Knowledge lecture)

In short, Peace lies within balance, not extremities. Balance can be found through moderation and diversity. Diversifying your resources before concluding an idea, or, choosing to surround yourself with all peoples, instead of isolating yourself or surrounding yourself with your clone, is how we can find peace and serenity. When we isolate ourselves or keep ourselves imprisoned in narrow-minded ideology, we are killing our potential.

Spiritual Focus (Exercise II)

How do we unlock our potential?

The subconscious can be both fascinating and frustrating at the same time. There are a time and a place for the information in this place, but when it is not the time to wander in the subconscious and super subconscious, we need to know how to stay focused. For this reason, it is important that we learn to stay focused when we meditate.

Focus meditation training consists of trying to focus your mind on a single thought, for as long as you can. The longer you can stay focused on a single thought, the easier it will be for you to meditation and conduct self-healing practices.

The training is simple:

Think of something, such as your houseplant, and try to ONLY think about that one thing. Try examining the entire structure and entity of that object.

Time yourself. Try to reach a minute, at first. Then, try for 5 minutes. And so on. In the beginning, you can start this training by staring at the item while you meditate. Eventually, you will want to move to rely completely on your own recorded memory. Once you feel confident that you can remain focused for at least 5 minutes, then you are ready to conduct your first journey to meet your Eternal Self.

Part III: ZERO

Spiritual ZPE

The concept of Zero refers to the exact measurement of negative energy and positive energy that they cancel each other out. Earlier, we learned about this Neutral Point, when we explored the concept of Black Matter and Black Energy, in which we discovered that both Black Energy and Black Matter are the equivalents of Zero-Point Radiation or Energy (ZPE). ZPE is the grounding energy that makes up the universe, which is a vacuum or void of energy that all life comes from and reverts back to when stars and solar systems die. This point of nothingness can be recreated from a spiritual perspective, too. In spirituality, the equivalent of ZPE is a mental state of clarity and infinite awareness, which is also referred to as the God State of Mind. In this God State, we are capable of conceiving infinite knowledge and wisdom, a power that stretches beyond a single moment in time or space... simply because we have become in sync with the essence of both time and space.

Space, by the way, does not only refer to the universe. Space can also refer to dimensions, realities, and environments. This means that if I say that an action rippled throughout time and space, I mean that this action affected other perspectives all around the world and to predecessor generations. For example, the Great Sphinx of Egypt was not only monumental in the time it was created but has inspired countless others who have both seen this monument with their own eyes or by word of mouth or through pictures for thousands upon thousands of years.

This monument affected the masses throughout time and space.

In this reality, Earth, it is very difficult to reach that level of awareness... or rather, maintain that level of awareness. The reason it is such a difficult and seemingly impossible task to do is that we are bound to the laws of this reality and the distractions associated with it.

It is an odd sense of irony that we need to exist in this reality to obtain the experience we need to advance in enlightenment; but these very experiences keep us too grounded, preventing us from reaching our full potential. Thus, trapped in a realm of distractions and emotional turmoil, most never reach the highest level of their awareness... except upon their death beds, when materials no longer matter. In those moments many become in tune with the energy of pure love. It is in those last moments when many finally understand, but it is too late. This is why it is important that we strive to achieve our potential and to understand our purpose. Our ancestors, at some point, understood this and left us a map to show us the way to Spiritual ZPE... the Tree of Life.

The map of the Tree of Life shows us how to ascend and to survive through different struggles. At the top of this tree, is the God state of Mind which is the equivalent of Spiritual ZPE. To climb up the tree, we not only have to survive the different scenarios of spiritual development, but we must also clear our chakra points to allow Life Energy to flow easily through us. This river of Life is, in fact, the snake of knowledge, that is ever spiraling upwards to our Crown Chakra point. In a healthy state of being, we should be aware and in control of our emotions, as well as being able to accept our identities. This profound self-understanding clears the internal pathway that allows this life energy snake spiral to our crowns, as it was marked by the crowns of the ancient pharaohs, which depicted a serpent leaving out of their third eye.

In this Spiritual ZPE mental state, we become masters of our awareness and have gained confidence in ourselves. We see beyond the initial issue and view even a glimpse of the greater picture. With both mental balance and confidence, we would be capable of doing amazing things, such as transferring knowledge from the cosmos to this reality – building great wonders and reaching beyond the laws of man to become innovative creators. A Sage is a person who has mastered Spiritual ZPE and has collected great amounts of wisdom. Famous Sages in history (whether real or rhetorical) are all seen as Buddhas, Abrahamic Prophets, Gurus and other Spiritual Guides of this era. I would even argue that individuals involved in the freedom and revolution of oppressed peoples, were also Sages. Though they may have not reached the highest level in understanding spiritual ascension, their actions depict that they were deeply in tuned to Black Energy, even if they were unconsciously aware. These unrecognized Sages were filled with the desire to push for justice, by any means necessary. In their unconscious God-like State of Minds, they spoke to people beyond space and time. Their words became prophetic and they moved mountains of people to react and produced action from those fixated in their boxes. These unrecognized Sages are: Malcolm X, Martin Luther King Jr., Marcus Garvey, Bob Marley, Ture Kwame (Stockley Carmichael), Dr Yosef Ben Jochannan, Dr John Henrik Clarke, Dr Welsing, Dr Ivan van Sertima, Amos Wilson, Nat Turner, Mammy Maroon, Queen Cubah, Freedom Fighters, Tacky, Peter Tosh, Selassie, Ana Mambi, Nina Simone, Lauryn Hill, and on and on it goes. In some cases, these Sages died at the peak of their awareness. In other cases, they became lost or went incognito. But regardless of their current status with the world, their words reached both our soul and spirit and continues to echo throughout time. This imprint that they left, will continue to become even more amplified as time continues on.

"The conscious must understand what their task is... ours is not to teach the people how to be conscious but to make them conscious of their unconscious behavior." – Stockley Carmichael (lecture The Unconscious vs the Conscious)

It is just so... the purpose of a Sage is to not teach the people how to be wise but to guide them to be enlightened in their own experiences. Malcolm X was a master at that. He was able to get his followers and auxiliary followers, to look deep within side themselves and to question their own reality. Wherever he went to speak, the people would begin to erect temples and organizations after being inspired by his words. He made the racist aggressive white man, stop and think about his actions. And by doing so, he forced that racist man to develop a conscious about his own behavior. That is power. A gift that goes beyond the standard thinking of an average man. These are words drawn from Black Energy at its most concentrated form of blackness. There are countless individuals that go unnamed today, that shoulder this burden of this power and go forth in the world with a determination to make it better. Individuals who are of color, who devote their life to rescue individuals from white supremacist organizations and introduce these souls to a world without hate. Individuals who are children that speak like adults, who inspire millions to want to save the planet. Individuals who can look deep into someone's heart and know that all they need is a simple act of affection, to change their entire perspective about the world.

I heard of a person who would go around smiling at people, as they went about their daily tasks. They made that extra gesture to smile at strangers. When asked why they said that they realized the power of a smile. A stranger may be on the verge of wanting to commit suicide and the smile they received that day may change everything. That action is the action of a Sage. Someone who thinks well beyond the moment and can see a larger picture and understands how they can affect the reality that they are in. Imagine if we were a society of Sages. In a society of Sages, we could build this world to be both productive and beautiful without harming the balance of this planet. As a society of Sages, we would never have to fear to come to the "end of the world" because we would have solved the problem and cured the world of egotism. We would live our lives striving for balance instead of materials. And we would each have inner peace.

Meanwhile in a society with an egotistical state of mind, in which people are focused on the moment or focused on Today while dismissing the importance of Tomorrow. This society of egotistical fools lives life imbalanced and destructive and unsatisfied. This endless thirst for trivial things is destructive not just to the planet but to ourselves. This act of changing our civilization of savagery to becoming a civilization of Sages is the only way we are able to save the world against egotism. It was inevitable, that the main purpose behind all of the movements from all of the past Sages was to strive to help us change our way of thinking from egotistical to revolutionary.

"Our job is to not teach people to be conscious, but to make them conscious of their unconscious behavior because unconsciously they seek freedom." – Stockley Carmichael (lecture The Unconscious vs the Conscious)

By now, you have discovered the patterns that unconsciously the unconscious will imitate cosmic order. A part of cosmic order is creative expression and freedom. Therefore, we are instinctively resistant to being oppressed. Therefore, this oppressive system goes against our nature. We fight to have the right to live in a better environment. We fight for the right to be acknowledged. We fight for the right to be treated equally. We fight for the right to be loved. If these things were unnecessary to our reality as humans, then we would have no desire to unanimously echo the same cries for justice, as they have been spoken throughout every era for every tyrant throughout the entire existence of humanity. Our understanding of the patterns of natural law is that if something is unnecessary for survival, then it dies off or disappears because it was deemed to be useless. And yet… the fight for freedom and justice is continuous. And, it is spoken by all of the Sages throughout history. Therefore, this desire for justice is necessary for survival, in accordance with the laws of both nature and the universe.

But what is justice and what is freedom?

It is a curious question to answer but it is important to understand that many views both justice and freedom differently. One example of freedom can be found domestically in the town called Slab City, California. This town is situated in the Sonoran Desert which is in the Imperial County within the California Badlands. The population is primarily white American. In this town, they do not abide by any laws. As a result, the town residents (a community of squatters) demonstrate no order, trash is thrown every, drugs are used openly. There is very little orchestrated civilization other than living within close proximity to each other. In a documentary about the town, a resident randomly broke the glass and explained to the camera crew that the residents break bottles and leave the broken glass on the ground, just cause they randomly feel like it. To the residents in this town, this is what "freedom" looks likes.

But is it a healthy and balanced view of freedom? No, it is not. It is an imbalanced warped view of freedom, in which the residents have convinced themselves that living in a self-destructive environment is normal. The media probably puts focus on the residents who heavily have used or are currently using drugs in their documentaries for this specific town. But in the tent cities that are popping up throughout the west coast of the United States, the residents of these tent cities display a similar destructive mindset, in which waste is not properly taken care of and that their primary view of freedom is heavily associated with using their drugs openly around other users.

A healthy desire for freedom and justice is associated with a perfect balance, where unjust actions meet an equal punishment and where people are able to live constructive healthy lives without being oppressed. Imagine living in a world where you are able to be the best version of yourself without being judged or put into a box. Or a place where you no longer need to be afraid and never want for any of your essential needs. This is the ideology of freedom that has been used to manipulate massive amounts of people to follow organized religion. In the religious order, this "milk and honey" world only exists in the afterlife. Using the carrot and the stick techniques, religious organizations present this dream as being only obtainable if followers submit to their ridiculous laws. As we have seen in Muslim terrorist bombing attacks, the bombers were motivated to do these horrific things because they were told that they would have 72 virgins to conquer in the afterlife.

It is amazing how powerful organized religion is, to convince a man to disregard the physical woman beside him for fictional ones in the afterlife that he has no proof exists. Organized religious has convinced millions (if not billions) that their life is literally meaningless. And has therefore removed the desire to try to make this dream of a balanced world, an actual reality.

We are told that global peace is impossible because of mankind's greedy desires. But do we actually have proof that we are incapable of developing discipline against primitive savagery?

There is no proof that supports the idea that we are incapable of creating global peace. The only proof that we have is that:

a) The number of the unconscious and undisciplined have historically outnumbered the number of Sages.
b) When the majority of society was mostly in tune with a Sage mentality, they created unimaginable wonders.
c) When a society that is primarily undisciplined is educated, this same society can become disciplined and potentially innovative.
d) Human thought processes are globally connected and can abruptly be changed when 11% of a population's mindset begins to shift. Indirectly, the remaining 89% unconsciously evolve or mentally shift to match the shift of the 11%
e) It is only in the 21st Century when the world has begun to regularly communicate and learn with each other. This has become possible through the World Wide Web and global trade. Since

this is a new playing field, the limits of what can be produced in this new reality are endless.

f) Due to the Internet and other Global Interactions, civilizations are becoming more and more affected by global events that do not affect them directly. <u>Example:</u> Natural and Unnatural Disasters almost automatically generate a movement of foreigners to want to assist. This concept of auxiliary followers to a viral extent is unique to the 21st Century.

In other words, we don't know until we really try. Our ancestors didn't know if they could build a hut until they laid the first stick down and began to build what they envisioned. They did not know that they could build carriages until they pulled out their hammers and constructed one. They did not know that they could build pyramids or statues that would last throughout the ages until they laid down the first brick or carved the first chunk out of the asphalt.

> *"No one is born with the Truth inside of them... the purpose of life is to learn how to identify a lie... If anyone says that they are born with the truth inside of them, then they are a liar."* –
> Stockley Carmichael.

Similarly, no one is born with complete instructions on life, inside of them. We are given life so that we may be able to discover these possibilities and evolve to become better than we were before. If a man (or a woman) claims that they were born perfect. My advice would be that you should run far away from that sociopath. If a man (or a woman) claims that they never need to study or research anything because everything they need to know is already inside of them, then my advice would be that you need to run far away from that superficial fool. As both the superficial fool and the sociopath are hazardous to your health.

Our responsibility is to protect this planet and the creatures in it, so that other souls have the opportunity to ascend. We are all someone else's ancestors. Therefore, to deny the *Importance of Tomorrow* and denying the *Importance of Learning from Mistakes*, is to deny your responsibility to your future and the future of futures. To live a life of selfishness is to condemn another's spirit to damn-nation, cursing them from achieving their own potential and setting them up to fail.

 As it is our responsibility as Sages to share knowledge with the world, so that we may all benefit from global advancement. It is our responsibility to make this world better on our way out than how we received it, on our way in.

 Unfortunately, the *Importance of Tomorrow* cannot be comprehended by those who are not aware of themselves and their potential. Sometimes life may seem lonely or useless. When we become stuck in our own torment, we may find ourselves leaving the task of "fixing things" onto our predecessors, while we live the rest of our life making excuses as to why we refuse to change.

This lie that states that we "are stuck in our ways" goes against nature. We are products of change which means we are forever capable of learning and evolving and adapting. The only limits are the limits we place upon ourselves. In this chaotic period, we need every able-bodied spirit to step up to the plate and play a turn, in order to change the outcome of our future.

You have a purpose. You are needed. You are valuable. You are surrounded by love, even if you cannot see it, yet. You are capable of loving and being loved. You are capable of mentoring others. You are capable of being a hero. You are capable of achieving Spiritual ZPE and capable of leaving a powerful impression on this world.

In short, Spiritual ZPE is the God-like state of mind that a person can potentially reach, which is where an individual becomes perfectly in sync with Black Energy. This state of mind can last throughout a lifetime or for only a moment. In this time, a person that reaches Spiritual ZPE is able to see beyond average moral constrictions and get a glimpse of the larger picture. During Spiritual ZPE, individuals have historically conspired great wonders and ideology that motivates and inspires legions of the unconscious throughout time and space. A Sage is a person that is regularly in sync with Black Energy. The actions or movements of a Sage are in line with cosmic order, which is to create balance, life, justice, and freedom from oppression. Everyone is capable of becoming a Sage. However, in this reality, we are constantly blinded and emotionally distracted in Chaos, which blocks most of us from reaching our true potential.

The Tower and Fictional Karma

Ever since we were little, we have been taught or naturally fell to the belief that an unknown power will protect us. This idea is a part of our hopes and belief. As we learned in the chapter Knowledge vs Believe, a belief has no proof to confirm its existence. In light of that fact, we have no proof to support the ideology that an unknown force is willing to come and save us from ourselves. This desire to wait for a higher power to rescue us from ourselves has led us to combine and connect our experiences together to formulate unhealthy reasoning or justification for horrific and traumatic acts.

For example, if a girl is raped by a grown man with AIDS because he believes in the superstitious lie that sleeping with virgin girls will heal him of the disease. Then it is justified as being labeled as a part of God's plan. When children die while watching vultures eat on their starved bodies, it is labeled as God's mercy. Or that when countless children died in the floods of New Orleans after Hurricane Katrina or in Haiti or Porto Rico because a president fears people of color, it is labeled as God's mercy. Or that when millions of slaves were drowned at sea, it was labeled as God's plan. Or countless of innocents were brutally slaughtered by the Crusaders, it was labeled as God's plan. If a child dies from Sudden Infant Death Syndrome (SIDS), then it is justified that their death was somehow intentional to God's plan. If the Indigenous people in the Amazon jungle are burned alive, then it is justified that it is a part of God's plan.

This method of ***pseudo-justification*** is a form of insanity, that has no limit to how low justified pain can be bargained off. Instead of crying out for justice or being motivated to fix the problem, the poor souls fall to their knees and embrace the hardship as being justified.

This Pseudo-Justification should be categorized as a mental syndrome, like Stockholm Syndrome or the Imposter Syndrome. As the behavior found in Pseudo-Justification Syndrome is a coping mechanism that enables us to be content and dismiss our internal turmoil instead of learning how to face it. This undeniably creates other health problems associated with increasing Cortisol Levels. (Cortisol is the chemical our body produces when it is under stress) In other words, ignoring your feelings will cause anxiety. By dismissing the need for you to react to a situation, you are creating a chain of emotional events that you will inevitably face down the road.

Years ago, I was brought to a funeral for a friend of a friend's child. This couple's daughter had died of SIDS. The father stood on the podium and he talked about his daughter and how beautiful she was. At the end of the speech, he sat down beside his wife, as the preacher took the stand. The preacher began to tell the couple and the audience how it was in God's plan that she was taken so young. That this child had already done everything that she needed to do in life, that God decided to take her from her family.

That moment forever haunted me. My eyes remained focused on the faces of this child's parents, who grieved heavily. I memorized the details that they put into making sure everything in this funeral was perfect because they loved that child so deeply. And I thought, how cruel it was for any God to do this to a loving family.

No matter how much I wrapped my mind around the words of the preacher, it didn't add up. It has been years and still, I wished that at some point after that funeral I would have told the preacher that **they were a lie**. How dare they deny this couple the right to be angry that a cure has not been found for SIDS! How dare they deny this couple for having the desire to fix the problem or to motivate others to do so! How dare they plant that nightmarish seed into the couple's minds that if ever they have another child, that God might randomly take them away, too! All for the sake of what? A mere temporary solution for the moment. It was like the preacher had slapped a tiny band-aid on a giant wound. Indirectly the preacher basically said to this couple that God wanted them to "get over it."

My father died a few years before writing this book. He died primarily due to poor service at a Floridan hospital. His emergency calls were ignored so that by the time they came to his room, he was dead. Knowing how black people are dismissed in the medical industry, I know that if it was different and if he was white, he would have not been ignored. I know this in my heart. But at his funeral, another preacher echoed the words of the other by dismissing his death as being a part of God's plan. While the rest of my family and siblings embraced the dream that his death was in God's plan. I made a promise to myself, which is that I refuse to let anyone dismiss my right to feel emotional. It was wrong, how he died and one day, I will take this energy and fix the system. And that's that. Does it make me feel better? *Hell yeah, it does.* I am motivated and inspired to do good things in my father's name. I have a choice to either be constructive with this energy that I have received from this situation or chose to be destructive. Regardless of what many choose to believe, inaction (being submissive to the pain) is an action and it is a destructive action.

The energy I sensed from the father of the child was so powerful. If pointed in the right direction, he would have been an organizer who would have moved mountains to find a solution. But in a blink of an eye, his power was diminished as he embraced God's will in hopes of obtaining a miracle.

 Miracles do exist, but not necessarily in the way that you think. There are cases of a phenomenon, where individuals have escaped near-death situations and experienced a form of miracle. Is it coincidence or was it intentional? I, myself, have had many miracles that have occurred throughout my life. In those situations, I do not doubt that these scenarios are miracles. But I also feel that there is more to the explanation than it being the will of an unknown deity. There are countless people have also prayed for asylum and have died horrible deaths. Organized religion plagues this planet, so it is very likely that the victims of these horrors also prayed faithfully to the same deities.

But they died, nevertheless.

Why do some die and why do some life? Could it be because of chance or it is possible that our ancestors did step in to help us or is it that we suddenly became in sync with universal frequencies to influence a change of events? Or is it a combination of everything, including a mystery deity rushing in to save us? It is a lovely concept to believe in divine intervention, but it is also equally foolish to bet your life on it.

Sometimes there are events that will happen that seem to be beyond our control. Leaving us fearful for the future or fearful of death. As a result, we pray to be saved or pray for Karma because we live by a moral code. But the real villains of the world, do not. They keep on harming us because they know we will remain in our boxes even if we are being eaten by vultures.

One of my favorite miracle stories is in the area called ADDIS ABABA in Ethiopia back in 2005. Where a 12-year old girl was kidnapped by 7 men, who were trying to force her into marriage. She screamed and cried for help. In this midst of this helpless situation, do you know who came to her rescue? Lions! Lions rushed to the girl's aid and chased off the kidnappers and protected her until a rescue party came and found her.

> "The girl, missing for a week, had been taken by seven men who wanted to force her to marry one of them... She was beaten repeatedly before she was found June 9 by police and relatives on the outskirts of Bita Genet... She had been guarded by the lions for about half a day...They [the lions] stood guard until we found her and then they just left her like a gift and went back into the forest," said

Sgt. Wondimu Wedajo (NBC News, Jun 21, 2005)

This miracle left many to wonder: *was its divine intervention or coincidence that the lions thought she was one of theirs or perhaps in the midst of her screams, she connected or synced to a frequency that called other beings into action?* Based on all that we have learned in this book, I would wager that it was the latter possibility, which we will learn more about in the chapter called **432Hz**.

A chaotic world is the essence of the tarot card called the Tower. It is a nightmarish time driven by many self-destructive situations. It is, by far, the worst card in the deck and the card that many fears, as it represents a world when things are going on beyond our control. I feel as though, this concept of having no control over something (or someone), is a scary thought, particularly for strictly dog lovers. When I listen to diehard dog lovers talk about why they love dogs, it is usually associated with them having complete control over the animal, which they label as the animal being loyal. These same individuals ironically despise cats. When asked why they hate cats, their reasons are associated with the ideology that they have no control over the cat and therefore cats are either assholes or scary.

Meanwhile, when you ask strictly cat-lover type people why they love cats, they usually state that it is because they can respect the animal's will for independence. They look beyond the idea that they "own" another living being. They find that they can relate to the animal because, like them, they have their own willpower. Or in other words, they come to a mutual understanding between the cat and the owner. They have no desire to have complete control or dominion over the animal, they instead establish a balance of space and life. In Buddhism, one of the hardest practices of meditation (which is the last stage of reaching harmony) is to sync to the mind of a cat. Additionally, the Kemetic people felt the same way, where the dog was portrayed as a servant. While the cat was worshiped as Powerful Goddesses (Sekhmet and Bast).

How this all ties into the Tower card is that those who fear to have no control (those with the mindset similar to dog-lovers) will find this situation to be the hardest to overcome. Meanwhile, those who look passed the face-value (those with the mindset similar to cat-lovers) will be more lenient to adapt to the situation and find it easy to survive. Every card has a clue as to how to resolve the situation. Therefore, the clue for this situation is in sync with another card called the Wheel of Fortune... *which is the very essence of this book.*

The solution for even the most nightmarish situations is that the bystander or the person praying for outside help actually holds the power to change the problem. Or the ability to easily adapt. By connecting to power found within your Eternal Self, you have access to a map to lead you through the situation. With a map, you can easily ride the waves through life regardless of how intense these waves coming at you will be.

How does this tie into fictional Karma?

One of the most difficult chakra points to clear is the mind's eye. This chakra point is blocked by illusion. To clear this point, the question we must ask ourselves is:

What illusion are you living in?

In the scenario of the Tower, the illusion is karma.

It is true that every action will lead to some form of consequence. But the consequences of the action are not necessarily justified. When Treyvon Martin was killed, his killer never was served this hypothetical karma. He has continued to live his life profiting off getting away with murdering an innocent boy. Years into the future, when this killer finally dies, we cannot say that his death was in relation to his cruelty. We have no proof that in the afterlife Treyvon's murderer will be punished. What we can physically see is that this murderer lived some form of life, while the bystanders cried out for someone else to serve them justice. The bystanders stood by a system that failed them and continued to stand by the system living on the belief that karma will be served.

It is a very pretty picture, this idea that horrible people receive just punishment for their crimes, while compassionate people need only be patient and wait for a higher power to step in. Unfortunately for me, I have stopped believing in the tooth fairy a long time ago. I know that when I put my tooth under my pillow, there will be no coin to replace it. What I know is that if I want a coin, I need to go out an earn it.

In the setting of the Tower, the character is endlessly hopeful for divine intervention. Similar to the mourning father, this character has two choices, which are:

    a) To take constructive action.
    b) To take destructive action.

In this time of chaos, we are given many gifts to help us through our decisions. We are given the gift of healing crystals, our ancestors, our orishas (guardian light beings) and our Eternal Self. We were given access to these resources so that we can enhance our lives. *As powerful as these beings are, it is not their responsibility to protect us from ourselves.* That is the key thing I have had to learn the hard way. If they are our ancestors, they have already lived their lives. If they are the orishas, they have many others they must inspire. If they are healing crystals, they are only temporary solutions until we can ascend on our own. And finally, if they are our Eternal Self, then they are confined to our core. It is up to us, to take to steps we need to take so that we may survive and to connect to a spiritual frequency.

432Hz

Fre·que·ncy (noun)

The rate of a vibration (which is measured by a wave) that happens either in soundwaves or in electromagnetic field (as in energy).

Frequency is everywhere and plays a major role in the morphogenesis process of this reality and everything in it. Frequency (whether sound or electromagnetic fields) are invisible strings that guide us or triggers that control our core genetic infrastructure to move specific elements and particles into alignment. To put it differently, the formation of the universe was orchestrated to a Contrabass, in which the low octaves (or low frequency) set the groundwork for other melodies to come into play; creating various layers to the final masterpiece. The concept of frequency has been studied for hundreds (if not thousands) of years. These studies of frequency have been conducted in different sects of science and spirituality. However, when we bring the data collected from these various studies together, we discover a similarity and a pattern.

Whether it is Quantum Mechanics, Physics, Spirit Science, Animal Communication, Linguistics Chem-E and even Music Technology; each ideology essentially says the same thing. The difference lies within their terminology.

The argument is not if frequency exists but rather, how frequency was able to build this complex place. There may be many hypotheses about this phenomenon. My theory is that the Contrabass not only triggered other particles and pieces to move but also in their movement, triggered them to adapt to their own frequency tones. This concept is stemmed from the data gathered from studying how hearing is formed in both animals and humans.

Firstly, it is important to note that the sound range in a healthy young adult is from 20Hz to 20,000Hz. Our ability to transmit sound is processed through the basilar membrane which is located in the Cochlea (in the inner ear). Inside the basilar membrane is sensory cells that are called Hair Cells – they are named hair cells because of their hair-like strands that cluster on their tops. These hair cells are not only spread across the basilar membrane, but they are arranged in a specific order. You see, as the hair cells are developed during our embryo phase, these cells begin to adapt so that they can each receive a specific frequency range. Along the basilar membrane, the hair cells are mysteriously arranged from lowest frequency sensors to highest frequency sensors. Science remains still unsure as to how the hair cells were able to form in this way. According to the National Institute of Health (NIH) *"a research team led by Drs. Zoe F. Mann and Matthew W. Kelley at NIH's National Institute on Deafness and Other Communication*

Disorders (NIDCD) suspected that a molecular concentration gradient may guide the cells during development. Like numbers on a ruler, the cell positions might be marked by levels of a signaling molecule."

In one study by Nature Communications (published on May 20, 2014), the project examined 6-day old chick embryos. The auditory sensory system, the basilar papilla, in chickens is similar to the cochlea found in mammals, in which hair cells are organized along the membrane according to the frequency they are dedicated too. At the time of the chick's basilar papilla formation, the scientists discovered that during the morphogenesis process there was a concentration gradient of bone morphogenetic protein 7 called **Bone Morphogenetic Protein 7 (Bmp7).** Bmp7 is a protein that is categorized under the **Transforming Growth Factor Beta (TGF-β)** family. **TGF-β** is found in a variety of species and is essential proteins that are fundamental to the development of basic biological functions.

Bmp7 was traced across the base of the basilar papilla membrane during its early stages of formation. This signaling protein is produced during the embryonic stage and plays a key role in the development of the kidneys and bones. What Nature Communications discovered was that *Bmp7 also* played a key role in the development of low-frequency-sensing hair cells. The scientists took their research a step further, in which they bathed a basilar papilla membrane (which was still at its early stages of development) in a solution of Bmp7. The result from doing this was that the developed hair cells that were devoted to high-frequency would adapt to match the low-frequency hair cells.

In accordance with the concept of the universe being built on a Contrabass, the reaction of the hair-cells adapting to detect lower frequencies after being exposed to high levels of Bmp7, supports the idea that low-frequency is fundamental to basic biological developments and would be the explanation behind lower frequency sensors being formed first as well as their alteration to high exposure to Bmp7.

In another study by Nature Communications, electric frequency was tested against a living cochlea (in a mouse's inner ear), the reticular lamina (which is a thin tissue that is on top of the cells in the inner ear) reacted , in which the tissue fluctuated then returned back to its original position when the frequency stopped. There were a few minors aftershock fluxes. In contrast, the basilar membrane (which resides at the base of the hair cells), also reacted but reacted in a delayed time. Where the reticular lamina reacted easily to all frequencies, the basilar membrane reacted primarily to lower frequencies in delay waves. The results of this research also support the concept that the basilar membrane (which was developed under high concentrations of Bmp7) is designed to receive lower frequencies not higher.

We can find examples of ELF throughout the universe, but the most solid research is based on Earth. The **Schumann Resonance (SR)** theory was created to study ELF of the Earth's electromagnetic fields is measured by recording spectrum peaks, which are calculated off lightning discharges. The ELF measurements for the planet ranged from 7.86 Hz to 8 Hz. At any given time, there are about 2,000 thunderstorms that are active throughout the planet. The SR research team used global research stations and magnetic coils to track and measure the Earth's ELF levels and activities. For every 2,000 thunderstorms, the SR research teams were able to track SR signals off of ~50 Lightning events per second. This led to the development of the following formula:

$$f_n = \frac{c}{2\pi a} \sqrt{n(n+1)}$$

a. f_n = represents the *resonant frequency*.

b. c = the speed of light

c. a = the Earth's radius

d. n = is the *n-th* power

Based on SR research, the project was able to determine that other planets also generate their own ELF patterns, particularly: Venus, Mars, Jupiter, Saturn, and Saturn's biggest moon Titan. In reference to the Universal Contrabass theory, in which the low frequency produced by Black Energy not only animated the particles and matter to develop into other structures; but these newly formed structures were able to produce their own frequencies (melodies) that fluctuate in harmony with the Universal Contrabass.

From this development, we can find additional connections with other activities on this planet.

Infrasound is another term used to describe extremely low frequency (ELF) on this planet. Infrasound is specifically categorized with being within the range of 20 Hz down to 0.001 Hz. This frequency range can be used to monitoring earthquakes, charting rock and petroleum formations below the earth, and also in ballistocardiography and seismocardiography to study the mechanics of the heart. In other words, Low frequencies track subsurface activities of the planet as well as subsurface activities in the human body. When scientists are tracking Natural disasters, they are able to do so with infrasonic arrays because Natural disasters produce infrasound. Another area we can find infrasound is in **Animal communication**. There are a handful of animals that particularly use infrasound to communicate. These animals are:

- Whales: 10 Hz to 31 kHz
- Elephants: 15 Hz to 35 Hz – elephant's infrasound can vibrate through solid ground and sensed by other land animals using their feet.
- Tigers: 18 Hz (used to paralyze their victims)
- Hippopotamuses: 14 - 24 kHz
- Rhinoceroses: 5 to 75 Hz
- Giraffes: 12 Hz
- Okapis: 12 Hz
- Alligators' growl: under 20 Hz
- House Cats 18 Hz to 150 Hz

Even many species of migrating birds use naturally generated infrasound, from sources such as turbulent airflow over mountain ranges, as a navigational aid. When analyzing these animals, we find a common pattern which is that the animals are ancient species that existed during the prehistoric period (or directly linked to ancient species) and are all linked to the continent of Africa (minus the Tiger). In light of this, we may be wanting to ask:

Is infrasound an Ancient animal language?

The answer to that question is yes. It is the only logical explanation, as the planet produces infrasound as well as prehistoric animals. Normal matter (the particles that reflect light – Gamma Photons) produces Extremely Low Frequency (ELF) in comparison with the rest of Space. In the sea of Black Matter, the energy waves are more hyperactive than that found in normal matter. According to K. Rajpal in his abstract titled Dark Protons and Dark Matter:

> "Dark photons may be the particles of the elusive dark matter... The photon frequency available is continuous and has no upper or lower bound. There is no finite lower limit or upper limit on the possible energy of a photon. However, going by observations, the most energetic photons are the cosmic gamma photons with a wavelength of 10^{-15}m. The least energetic are the Very Low Frequency (VLF) radio photons with a wavelength of 10^5m...
>
> [Normal Matter] PHOTONS – classification as per their wavelengths:
>
> - Radio photons 10^5 to 10^{-1}m.
> - Microwave photons 10^{-1} to 10^{-3}m.
> - Infrared photons 10^{-3} to 10^{-6}m.
> - Optical photons 10^{-6} to 10^{-7}m.

- Ultraviolet photons 10^{-7} to 10^{-9}m.
- X-ray photons 10^{-9} to 10^{-11}m.
- Gamma photons 10^{-11} to 10^{-13}m.

Photons with frequency or energy more than gamma and less than Planck photons can be called **dark photons**. These may be the particles of the elusive dark matter."

In the chapter titled ***Spiritual Evolution Theory***, we explored the possibility of how time is reflected on all different levels of reality. By that theory and the report that Black Matter is more hyperactive while Normal Matter is more slowed down. Therefore, we could probably conclude that Spacetime varies upon the speed of frequency.

Concerning our review of Creative Energy, this discovery that black energy is hyperactive further supports the ideology that creative energy and black energy are the same. In studies where they examined the brain activity of artists, they discovered an increase of grey and white matter (which are composed of nerve cells) in the cerebellum and other areas of the logical brain that is tied to supplementary motor activity.

"The people who are better at drawing seem to have more developed structures in regions of the brain that control for fine motor performance and what we call procedural memory," Rebecca Chamberlain from KU Leuven (BBC Radio: *Artists 'have structurally different brains'*). "...This region is involved in a range of functions but potentially in things that could be linked to creativity, like visual imagery - being able to manipulate visual images in your brain, combine them and deconstruct them... Put to rest the facile claims that artists use 'the right side of their brain' given that increased grey and white matter were found in the art group in both left and right structures of the brain."

This increase in neurological activity could be the evidence needed to further understand how creative energy flows within us. This development of increased activity also supports another theory, which is the **Universal Contrabass**.

The concept of the Universal Contrabass is that there is the main frequency (whether sound or energy) that animates other items into action. This animated particles and beings, eventually formulate their melody frequency that is in harmony with the original tone. This main frequency would essentially be a **binary system** (at this point, I am sure you are seeing a pattern in binary coding).

There is another theory that will correlate with the concept of the Universal Contrabass, which further explains the binary tone. This concept is called the Scientific Pitch Theory. The Scientific Pitch is also known as philosophical pitch, Sauveur pitch or Verdi tuning and was created in the 19th century. It is a concert pitching standard which is based on middle C (C4) being set to 256 Hz rather than 261.62 Hz, making it approximately 37.6 cents lower than the common A440 pitch standard. All the octaves (in the Scientific Pitch) of "C" and are an exact round number in the binary system when expressed in hertz (Hz). The octaves of C remains a whole number in Hz down to 1 Hz (with the silent 0) in the binary system. The formula for the Scientific Pitch is: $\boldsymbol{A_4 = 432\ Hz = 2^4 \times 3^3}$

How we were able to get 432Hz is that the 256 is a power of 2. With that said, the Universal Contrabass is 432Hz.

Black People and the Universe Contrabass

"I have always noticed that Afakans, when listening to music in 4/4 time signature, always clap on the 2 and 4; whereas non-Afrakans clap on the 1 and 3. So, it the music has a Boom-pap-Boom-Boom-pap rhythm; the Afrakans will clap on the pap (2 and 4) whereas the non-Afrakans clap on the Boom (1 and 3)." by Yuya Assaan-ANU (Grasping the Root of Divine Power)

Concerning melanin (a substance we learned about earlier that transports information throughout the nervous system) that is found in black people, we are lead to the assumption that black people are unconsciously in sync with the Universal Contrabass. Hence, their natural behavior with getting in rhythm with the music to the bass beat as well as their natural ability to be overwhelmingly creative. It would also suggest that genetically black people have highly developed areas of the brain that are responsible for fine motor performance (visual and communication), according to the study from Chamberlain. This hyperactive developed brain would also support the historical references of the evolution of homo sapiens in Africa, as their direct descendants demonstrate similar cognitive abilities. Furthermore, this concept of a hyperactive developed brain would further explain the formation of language (which is stemmed from Bantu sounds) as well as explain the study regarding telepathic communication found with the Aborigines.

In an Interview with an Aboriginal Woman, hosted by Bonnie McLean. The Aboriginal woman, Loraine Mafi Williams, states: "There is telepathic communication among our people, but it occurs between individuals, not in groups. Not everyone is good at it. The elders, medicine people, teachers have been trained in telepathy, but it is a skill that does require training and takes time to learn. A runner ("scout" is an Indian word, isn't it?) may indeed communicate back to certain individuals in a clan."

Is telepathic communication the result of being in sync with 432 Hz?

The root words in human language are sourced from sounds. These core tones and sounds can be traced back to a prehistoric language that is very similar to Bantu and Indo-European. As the prehistoric tribes migrated around the world and mingled with neighboring tribes, and as humans evolved, the language became more complex. With that said, we can also conclude that these prehistoric civilizations imitated how the prehistoric animals communicated in infrasound. Some humans can speak in tones that are under 20 Hz. There is a rumor (though no specific proof can be easily identified) that there are prehistoric tunnels and caves that are designed to amplify infrasound, like an ancient speakerphone. When higher frequencies are used, the sound does not travel. I do believe this to be true, even if I cannot find the evidence needed. Under this oppressive system, discoveries that disprove its dogma are destroyed or buried, such as the Kemetic and Phoenician scriptures uncovered by Ivan van Sertima in the Caribbean or the ancient pyramids found in the Caribbean ocean. If you can uncover more evidence about these buried discoveries, please share it with all of us.

Visions

By further understanding this Universal Contrabass, and the role it plays in animating creation. There is another theory that we should eventually analyze, which is the explanation of visions or dreams that come true. Are vision visuals that are played when we become in sync with binary frequency or infrasound?

When we study Animals' instinctive behavior to natural disasters, we are left to wonder if the animals are responding because they are in sync with 432 Hz, or they are reacting because they sense the infrasound with their hooves, paws, and feet. This study is a difficult concept to calculate because we are incapable of tapping into the thought processes of animals. We can only hypothesis their experience, in this matter.

Negative Frequency

We examined 432 Hz but what about 440 Hz...

There are a few sound frequency water experiments (where the pattern of frequency is examined in the ripples of water), throughout the internet and books, etc. In this experiment, we learn that the pattern differs between 440 Hz and 432 Hz. In one popular experiment, the individual tested both frequencies separately and then together. According to the researcher:

"This video is a Cymatics version of Sound Frequencies in Water: A=440 Hz vs. A=432 Hz. It has been said that music tuned in A=432 Hz is more beautiful and harmonious to the ears and induces a more inward experience where music tuned in A=440 Hz is more outward, a mental experience which is projected outwards. I am sure you have already seen the following still photos of how the A=440 Hz vs A=432 Hz looks like. They have been circulating the net for some time. Well, I thought it will be cool to generate them using my Cymatoscope…and so I did…in real-time! It is said that a picture is worth a thousand words… Can you see the difference? Yes, there is none as the frequency difference is only 8 Hz." - George Of the LA Jungle (Sound Frequencies in Water: A=440 Hz vs. A=432 Hz; Sept 15, 2014)

In response to the video, many found the 440 Hz and the sound combination of both 432 Hz and 440 Hz to be disruptive, where many stated that they were starting to feel sick, anxious and nauseated. Meanwhile, the same various commenters found that 432 Hz was calming and soothing. In short, the image may not appear to be different, but it is more about how our bodies react to the sound that determines if it is negative or positive. If there is a massive movement to switching music over to 432 Hz and countless people prefer listening to 432 Hz. Then, we can only assume that with so many experiences saying the same thing, that there must be more of a difference than simply 8 Hz.

Try it! The only one who will understand if 440 Hz is better or worse for you, is you. Try experimenting with different frequencies on your own, to see what makes you feel better or worse. Write down the results of your experiment and share it. Also, get your friends to do the same thing and compare notes.

Crystal Science

In the second section, we discovered the Power of Intent and everything associated with life energy. This includes the exploration of Auras and how they affect us. Auras are layers of energy fields that surround their energy source, which resonates from within our bodies. On a larger scale, the equivalent of Auras can also be found on our planet. Where the various layers of the atmosphere operate in a similar function, where the atmosphere surrounds an energy source (as the earth has its core energy). A healthy atmosphere aids our planet to be equally healthy. On an even larger scale, we have also learned that our solar system has a form of energy field that is surrounded by black energy and black matter. In the core of our solar system, lies a star which is a source of cosmic energy for life in this solar system. Just like the atmosphere of this planet, the rings that surround our solar system are the overall reflection of the health of this reality. And on an even larger scale, we can see the same ring pattern that surrounds our galaxy. As above, so below.

What does this have to do with crystals?

It is speculated that crystals are gifts that were bestowed to us by our guardians to help us ascend. The reason why the idea that these objects are gifts, is based upon how we react to these items. Various beings and elements seem to react to being near or exposed to a healing crystal. One example of this can be found when researchers tested out how water reacts to crystals.

They found that snowflakes that were formed from crystal water (water that was filtered with crystal stones) formulated beautiful patterns. While water that was not exposed to crystals looked drastically different and ugly in comparison. Another example could be found in how plants also interact with crystals, as crystals seem to enhance how well the plants grow versus without. As we learned in section one, plants are very sensitive to frequency and negative energy.

As are we.

On a less scientific explanation, the reason why crystals are important to us is simply that they make us feel better. When we meditate, they help us amplify our frequency as well as clarify the distortion in our Auras, which helps us to connect to our Eternal Self or Black Energy. They radiate in our melanin. They maintain the balance of our environment and our pineal gland reacts to them. Therefore, they help us with being able to become a better version of ourselves. Crystals are used in everything to transmit frequency waves, from electricity to sound waves. Crystals are found in every technological device and are a mandatory substance to make any technical function. This is due to the chemical structure of crystals, which is perfect and able to amplify and transmit frequency waves in a similar way that our pineal gland is capable of doing. Since we are organic antennas, just like how computers need Coltan crystals to function, we need crystals (or Pineal Gland – our embedded equivalent to crystals) to properly function and process frequency.

Try it! During your daily **5 to 15-minute** meditation, try surrounding yourself with crystals to help maintain balance in your Asé, Chakra points and Auras. Write down your experience, afterwards, to see if your meditation was enhanced.

Reiki Meditation Exercise (III)

What is Reiki?

Reiki is a Japanese technique for reducing stress and promoting relaxation that also promotes healing. It is the concept that is based on how Asé or Ki flows through us. The word Reiki is formed from the combination of two Japanese words - Rei which means "Higher Power" and Ki which means "life force energy". Thus, the proper definition of Reiki is "a spiritually guided life force energy." The practice consists of manipulating Asé to radiate off specific regions of the body so that healing practices can be performed. This practice is associated with Spiritual ZPE because you need to be in a Spiritual ZPE to control the flow of Asé.

Reiki Exercise for Self-healing.

Here is a Reiki Self-Healing practice and exercise. We are being by sitting comfortably in a chair with your feet flat on the floor so that we are grounded. Place your hands above your Heart Chakra point or cross your arms across the Heart Chakra point.

Next, we speak (5) affirmations out loud. Use the blank lines below to write out your (5) affirmations of choice:

(1) _____
(2) _____
(3) _____
(4) _____
(5) _____

After speaking those affirmations, you can lower your arms and focus on the direction your Asé is flowing. Try to focus the flow of Asé to a specific region, like the palms of your hands. The more you focus, the more the temperature on your palms should increase. This heat that is generated can be used to heal swollen or painful regions on your body.

Kemetic Focus Exercise (IV)

There are meditation and yoga practices that are found on the walls of Kemet.

These meditations are based on the belief system popularly associated with the cultures of Kemet, in which the user focuses on invoking the associations and powers of the deities Ausar and Auset. In other words, the user focuses on the characteristics and contemplates on how to connect with them. For example, focus on the story of Ausar and how he was divided than reconnected. How can you spiritually rebuild yourself to be a better version than you are at this moment?

Part IV: Tarot Cards

Learning the Language

Finally, we have reached the last section of this book. At this point, you should have a better understanding of the different levels of self. With a deeper understanding of the coding that makes you who you are, we can now begin to construct a language to communicate with our Eternal Selves.

A traditional tarot deck consists of a total of 78 cards. The standard deck is divided up into two parts, which are the Major and the Minor Arcana. Major Arcana tells the story of spiritual/human development and is the main story of the tarot deck. The Major Arcana consists of 22 cards which represent the 22 spiritual paths or principles found in the Tree of Life.

In the Earthly realm at the root of the Tree, we begin this story as *the Fool*. This character is naïve to the world around him. Eventually, as we progress through the 22 principles (or characteristics), we end up with the World that resides beyond the realm of Fire, which represents the symbol of completeness and gaining a new level of understanding.

In a deck of Tarots, there also is the Minor Arcana, which consists of 56 cards. These cards are the supporting characteristics and stories, to help tell and personalize each reading. The Minor Arcana is divided into 4 suits. Each suit is dedicated to an Elder God: Earth, Air, Water, and Fire.

There are many different types of tarots, each which have been designed to tell the same story. Other teachers of the Tarot, will teach using these cards as a means to communicate with the dead or mysterious spirits, etc. I do not teach that. I teach that the Tarots is a medium the consists of both images and the general descriptions that work perfectly as a method to communicate with your Eternal Self. The more you use these images, the easier it will be for you to understand the messages that are being sent to you. Therefore, there are so many different versions of the same story, as there are countless different perspectives that vary upon culture and race and environment and religion. For example, in the Aleister Crowley Thoth deck, Aleister Crowley changes the Major Arcana World name to Universe. As he felt it better depicted the ultimate result of this story, which is to have complete balance and to be able to have the power of the Universe, at your fingertips.

In the Voodoo Deck, the World card is depicted as a serpent's egg, on the verge of hatching. This symbolizes being born into a new stage of awareness. While in the Afro-Brazilian deck, the

World card is symbolized by a person handing out their offering to the world, while being surrounded by all the elements (or elder gods). The offering represents the person being thankful for receiving this new understanding. In the BLK Excellence deck, the World card is depicted with a person cloaked with black energy, sitting on top of the world, while they are interwoven with the Universe. This represents the person is aware, balanced and grounded... the ultimate state of being.

The characteristics and illustrations of the card are just one aspect of Tarots. There are symbols and astrology that is embedded into each card, to help the reader understand the message.

BLK Excellence Reading

There are different styles of tarot card reading, as explained earlier. For the reading style of BLK Excellence, the use of reading the cards in reverse is *not* necessary. The Major Arcana section goes over the characteristics, elements, equivalents, and advice for the card. The Minor Arcana section goes over the characteristics and advice. Also, in the Minor Arcana section, there are Positive and Negative characteristics about the card, to help you fully understand the energy's essence. I use the two out of three popular Tarot decks (*which are:* Louis Martinié's New Orleans Voodoo Tarot Deck Aleister Crowley's Thot Deck) as comparisons. I do not use the traditional Medieval Deck as a comparison because the BLK

Excellence deck matches the traditional deck layout. With that said, the reason why Martinié's and Crowley's decks are used as a comparison is that these two decks have a slight alteration in how they are structured. Crowley's deck, for example, changes the Minor Arcana Pages into Princesses. While Martinié's deck has an additional wild card (Les Barons) that is made specifically for voodoo card reading. Furthermore, the titles in the Major Arcana cards were also changed in these comparative decks.

Regardless of these changes of titles, the Energy surrounding the cards is the same. It is good practice to look at the card decks comparatively, to get a better understanding of the function of the cards. Some decks will have more symbolism associated with it, like Crowley's. While other decks will focus more on depicting the overall feeling of the card, like Martinié's.

After the sections that explain the cards, there are spreads that you can use to perform your reading. This is not the limit of what you can do with your tarot cards. Beyond the readings, I would highly encourage you to set up your Astro Natal Chart, in which you see the planetary placements at the time you were born. Once you create your natal chart, you can begin placing the tarot cards that are in correlation with the zodiac symbols on your natal chart. This will help you further understand your characteristics and your story. It will also help you understand how the cards work with each other.

Major Arcana

0/Fool

The Spiritual Characteristic of this card tells the story of a fresh new awareness of the world. The fool is a character, who sets out on his grand adventure without taking a map, with very little supplies and a few coins in his pocket for the trip. He chooses to trust in his impulse decisions and be-lie-ves that his decisions made in ignorance and naivety, will lead him in the right direction and that everything will be fine. While the fool is carefree, signs are surrounding him suggesting potential danger around him and ahead. Regardless of the signs, he maintains a carefree spirit as he runs along his path and ignores the excessive stumbles, dusts himself off and keeps on trudging forward.

BLK Excellence Deck depicts the Fool running up a slope. The terrain around him is ridged and rough. He is dressed in Kemetic garments that suggest that he is royal, demonstrating his potential.

- **Meaning**: New Beginning, carefree, optimism, trust and nativity.
- **Element**: Earth/Air
- **Energy**: Masculinity
- **Advice**: Notice the signs around you and learn what those caution signs mean, that way you can prevent things that may do you harm.

Tarot Deck Equivalents:

- World Egg (Louis Martinié's New Orleans Voodoo Tarot Deck)
- The Fool (Aleister Crowley's Thot Deck)

I/Magician

The Spiritual Characteristic of this card surrounds the energy of self-improvement and self-development. The Magician is the devoted student of his craft, who seeks to experiment with his ideas and make his theories into a reality. The magician is a young wizard or a young student, experimenting with the skills that he has gained on his journey when he was a fool. This character is more introverted as he tries out different combinations of theories, sciences, and concepts. The goal of the magician is that the character seeks to gain confidence in the new skill or new trade. His character represents this stage in our lives when we have had the determination to master a new skill or trade and have chosen to focus on self-development. In the BLK Excellence Deck, the image depicts a man carefully studying and experimenting with different elements. The character is very focused on building his skill.

- **Meaning**: Manifestation, Personal Power, Development, Experimentation, Creativity
- **Element**: Air
- **Energy**: Masculine

- **Advice**: The more energy you put towards constructing something, the quicker you will become a master of that skill. Focus and Patience are needed in this stage.

Tarot Deck Equivalents:

- Dr. John (Louis Martinié's New Orleans Voodoo Tarot Deck)
- The Magus (Aleister Crowley's Thot Deck)

II/High Priestess

The High Priestess is the next stage following the Magician. The Spiritual Characteristic surrounding this card is the energy of obsessive self-repair. Once the Priestess has gained confidence in her newly cultivated skills, she has also gained more awareness of herself and the Divine. In this phase, the high priestess takes time to clear out all of the energy, that she seems to be negative and believes blocks her potential power. She focuses on cleansing her spirit and life of negativity and to prepare for the future. The High Priestess is also a guide, who uses the new skills to help guide those seeking guidance and enlightenment. Her character represents our inner High Priestess, and she reminds us that it is okay to take time to plan and prepare and to cleanse ourselves for future endeavors.

BLK Excellence Deck depicts the High Priestess sweeping rapidly around her. She is not focused on anything else, except cleansing her environment. She doesn't even notice that her fanatic sweeping has been lead outside, where she is surrounded by endless work.

- **Meaning**: Stagnant, inactive, preparation, intuition, self-repair
- **Element**: Water
- **Energy**: Feminine
- **Advice**: During the frenzy of cleansing your energy, be careful not to sweep out those that may help you build your future.

Tarot Deck Equivalents:

- Marie Laveau (Louis Martinié's New Orleans Voodoo Tarot Deck)
- The Priestess (Aleister Crowley's Thot Deck)

III/Empress

The Empress is relaxed, laid-back and filled with an abundance of creative energy. The Empress, though relaxed, is filled with an active mind that is constantly thinking about new ideas and new dreams. Even though her mind is constantly racing with new aspirations, she still dominates her realm by ruling with justice, compassion, and authority. Her greatest characteristic is her empathetic nature; as she is nurturing. Her character reminds us that we need to recognize the needs that surround and/or affects us.

- **Meaning**: Creativity, fertility, abundance, intuition and nurturing
- **Element**: Earth
- **Energy**: Feminine
- **Advice**: Your creative energy is immense but doesn't allow your dreams to cloud your judgment and make you lose focus.

Tarot Deck Equivalents:

- Ayizan (Louis Martinié's New Orleans Voodoo Tarot Deck)
- The Empress (Aleister Crowley's Thot Deck)

IV/Emperor

The Emperor is confident, courageous, grounded with authority and wisdom. He has a clear idea of what he wants; or a clear objective. He is laid back because he is confident and is fully aware of his actions and the consequences of his actions, unlike the fool. Since he can understand the consequences of his actions, he has created stability for himself and those he has taken under the wing. In his heart, he is still rebellious and adventurous; he goes against the actions of the majority and finds his path, which is how he was able to find stability and confidence. He prefers to make the rules, rather than follow someone else's.

- **Meaning**: Structure, stability, dictates, power
- **Element**: Fire
- **Energy**: Masculine
- **Advice**: Avoid becoming lost in your ambitions to where you only prefer to listen to your council. Others carry great wisdom that you may need to ascend.

Tarot Deck Equivalents:

- Loco (Louis Martinié's New Orleans Voodoo Tarot Deck)
- The Emperor (Aleister Crowley's Thot Deck)

V/Hierophant

We found confidence as Emperors; next, we revert to the institutions that you have just broken away from. Yes, the Spiritual Characteristic of the Hierophant represents a character returning to the institutions of society and social tradition they previously broke away from, so that they can take another look. It seems a little, crazy... right? By returning to these institutions, you can now look at these systems with a fresh new perspective. With a fresh new perspective, you can see how these institutions influence your reality. The institutions that the Hierophant reviews can be categorized as education, religion, corporate culture, patriarchy, government systems, consumerism, family, friendships, love, and so on. Another way to look at it is that this characteristic returns to its roots to see if they feel the same or if they changed. This is a very inquisitive stage in human mental and spiritual development.

- **Meaning**: Institutions, tradition society, wise man, equilibrium, protection
- **Element:** Earth
- **Energy:** Feminine
- **Advice**: It is powerful to go back to analyze but do not become trapped back into the institutions you are trying to break free from.

Tarot Deck Equivalents:

- Master of the Head (Louis Martinié's New Orleans Voodoo Tarot Deck)
- The Hierophant (Aleister Crowley's Thot Deck)

VI/Lovers

The lovers are focused on embracing the bliss of their union. The spiritual character of this card represents a character who is experiencing a whole different realm and vision of life, that they didn't see before. Traditionally, the card depicts a man who is faced with two options (which is to pick a virgin or a prostitute). In the BLK Excellence deck, there are four characters in the picture. Two are seemingly innocent and faced beside each other, as a potential match. But their eyes are focused on two other characters who seem to be unpredictable. Despite any obstacles, the experience the characters choose is so exciting and blissful that the world may seem to cease to exist. This card isn't just about couples, but it can also represent the love for a hobby, interest, career and so on. As blissful as this stage may seem, there is a large amount of caution. This stage represents the dilemma and indecision between deciding to focus on Stability or Excitement. It also represents a phase of blissful unconsciousness; or carefree behavior that makes the character unaware of whatever else is going on.

- **Meaning**: passion, companionship, falling in love, bliss, choice, unity, and bonding.
- **Element**: Air
- **Energy**: Masculine
- **Advice**: Don't let your excitement and passion for something cloud your judgment or it may lead you down the path of chaos.

Tarot Deck Equivalents:

- Marassa (Louis Martinié's New Orleans Voodoo Tarot Deck)
- The Lovers (Aleister Crowley's Thot Deck)

VII/Chariot

The Chariot represents the spiritual characteristic best described as being at the crossroads. In this stage, we are preparing to move onward and upward, while defeating all the obstacles that stand in your way. However, with this mindset to push on, we are unable to move because we are internally conflicted. When we were thinking like lovers, we were careless and unconscious of our surroundings. In this stage, we have regained focus and determination to achieve success. Where the Lovers phase can mean the quest of an individual, the Chariot refers to Duality. This phase the struggle is highlighted as an imbalance between masculine and feminine energy, and the constant struggle to get both energies in order. To put it differently, we are internally conflicted with either moving forward in a practical direction or taking a more scenic route.

- **Meaning**: Duality, imbalance, internal conflict, overcoming obstacles, travel, dynamism, and success.
- **Element**: Water
- **Energy**: Duality
- **Advice**: Find a balance between both the feminine and masculine aspects about

yourself, to move forward. There is always another option that incorporates both energies.

Tarot Deck Equivalents:

- Dance (Louis Martinié's New Orleans Voodoo Tarot Deck)
- The Chariot (Aleister Crowley's Thot Deck)

VIII/Strength

This strength is the mastery of the currents of energy. There is no evil and there is no good there is only negative and positive energy. These energies surround us in waves of frequencies. We attract negative and positive based on our perception, actions, and thought. We can choose to reject the energy that comes our way or embraces it. Strength comes from the awareness of the power that can deflect and deflate negative energy. When we go against the current, the negative effects amplify, and we drown. Thus, the traditional symbol of Strength is of a young girl taming a wild beast with acceptance, compassion, patience, and understanding when and where her power should be implemented.

- **Meaning**: Patience, the right use of power, conquering fear, courage and primitive instinct
- **Element**: Fire
- **Energy**: Masculine

- **Advice**: Avoid rejecting the events that are happening, learn to accept these events as-is.

Tarot Deck Equivalents:

- Possession (Louis Martinié's New Orleans Voodoo Tarot Deck)
- Adjustment (Aleister Crowley's Thot Deck)

IX/Hermit

The Hermit is the mature and wiser version of the Fool. Even as an introvert, he still is an adventurer and a seeker of something new; however instead of focusing on the world surrounding him, he focuses on exploring all corners of his consciousness. He is unraveling the very essence of his being, by analyzing his thoughts, feelings, ego, and spirit. In a way, it is like the High Priestess, without the need to do excessive cleansing. He is hyper-inquisitive and questions everything and chooses to act based on his newly discovered truth. Consequently, the truths uncovered by the Hermit will create a trend of followers, who will follow the Hermit's ques and follow in the Hermit's footsteps.

- **Meaning**: Meditation, Awareness of Self, Quarantine, Isolation, Solitude
- **Element**: Earth
- **Energy**: Feminine

- **Advice**: There is power in solitude but do not make solitude indefinite as it will backfire and warp your reality.

Tarot Deck Equivalents:

- Couche (Louis Martinié's New Orleans Voodoo Tarot Deck)
- The Hermit (Aleister Crowley's Thot Deck)

X/Wheel of Fortune

The Wheel of Fortune represents the cycle of change in life and ironically represents the spiritual character of this book! This is the energy that was tamed by the maiden in the Strength card. Life is continuously filled with ups and downs. These changes can sometimes be scary, while other times it feels exciting. Spiritual Balance is found when you rest on the top of the wheel and have cosmic clarity. Inside the wheel, is the Eternal Self (which sits on the source of our Divine blackness). The seat of our inner self is the only part of the wheel that doesn't spin or go up and down. Your Eternal Self stays stagnant and observant of the events happening around it. Thus, power lies within being able to ground yourself by reconnecting or communicating with your Eternal Self. The BLK Excellence Wheel of Fortune card, shows a character meditating and outlines the different Chakra points in Kemetic and Ifa names. At the base of the character is the Earth/Grounding point, which is done by using a black stone (or by Grounding yourself). Each Chakra point is a different color. If you have crystals, the best practice is to use crystals that match the Chakra points.

- **Meaning**: Change, up and downs, need to find the spiritual center point
- **Element**: Fire
- **Energy**: Masculine
- **Advice**: When life is getting uncertain, imagine yourself like the wheel of fortune. Instead of fighting against situations, go with the flow and ride the waves. While you ride these waves of life, take time to reconnect with your Eternal Self. This will help you regain clarity.

Tarot Deck Equivalents:

- The Market (Louis Martinié's New Orleans Voodoo Tarot Deck)
- Fortune (Aleister Crowley's Thot Deck)

XI/Justice

The Justice card represents the Balance. The concept of what it just is best described in the 42 laws of Ma'at and the 42 Negative Confessions (see: Glossary). To understand what it means to be balanced, it is encouraged that we memorize these laws of Ma'at and the Negative Confessions. Justice is depicted with understanding the balance. This stage is about learning how to master rational thinking and be able to maintain balanced energy in your life. This is a difficult task as the reality that surrounding is always out of balance. There are layers to these thoughts and ideas, just as there are layers of meaning to this character. Therefore, to master justice means to keep your conscious clear.

- **Meaning**: Rational, equity, neutrality, just, balance, compassion, and empathy
- **Element**: Air
- **Energy**: Feminine
- **Advice**: Strive to maintain balance and positive energy will surround you.

Tarot Deck Equivalents:

- Secret Societies (Louis Martinié's New Orleans Voodoo Tarot Deck)
- Lust (Aleister Crowley's Thot Deck)

XII/The Hanged Man

The Hanged Man has not been placed in his predicament, forcefully. He has chosen to lazily hang upside down to see the world from a different perspective. His spiritual characteristic represents that period when we casually wait to see what will happen, instead of making a move. He represents the phase in our lives when we feel perplexed and need to take a moment to re-evaluate the situation. We may not be "forced" in our predicament, but we may be finding it very difficult to make a move. During this time of waiting to find new enlightenment, the character also awaits the potential judgment of his inaction. As there is no easy way for the character to regain control until the situation has passed. Meanwhile, most beings around him continue to go about their lives. But as he patiently waits, it would seem that some beings are waiting for him to make a move. It is hard to say if these bystanders are friend or foe. Either way, their fate is tied to this character's next decision.

- **Meaning**: Uncertainty, temporary situation, sacrifices, the complex dilemma
- **Element**: Water
- **Energy**: Feminine
- **Advice**: Sometimes being inactive about a situation is the only way to resolve it, however, be wary of an unforeseen consequence resulting from your inaction.

Tarot Deck Equivalents:

- Zombi (Louis Martinié's New Orleans Voodoo Tarot Deck)
- The Hanged Man (Aleister Crowley's Thot Deck)

XIII/Death

Death does not mark the death of a person but rather it marks the end of a situation, phase, relationship or experience. It is the marking point of a change that is about to happen. Just as Ra's death symbolizes new life and new stages in his life, this character represents something. The scythe or reaper cuts through the cords of experiences that a ready to begin dwelling in the past. Therefore, this card is liberating, as it represents the cutting of ties that are holding us back and letting us know that it is time to let go and embrace the next adventure on the horizon. When we let go of things that will no longer grow with us, we are free to blossom and advance into future experiences and adventures.

- **Meaning**: the mark of a change, a new direction, a new experience
- **Element**: Water
- **Energy**: Feminine
- **Advice**: Avoid trying to hold on to an experience that is becoming dead weight. Let go of these weights, so that you may be able to move on with ease.

Tarot Deck Equivalents:

- Les Morts (Louis Martinié's New Orleans Voodoo Tarot Deck)
- Death (Aleister Crowley's Thot Deck)

XIV/Temperance

This stage represents the formation the spirit must go through to be cleansed and structured and balanced. The process of forging metal into beautiful steel or heating raw gems to enhance their truest sparkle. We must gather the ingredients that we need to be able to develop ourselves to become our truest potential. This is a key point in our development when the power of introscoping first begins to play a more prominent role.

- **Meaning**: Balance, patience, reflection, duration, moderation, being sensible
- **Element**: Fire
- **Energy**: Masculinity
- **Advice**: Gather sustainable resources that will help you to reach your truest potential.

Tarot Deck Equivalents:

- Ti Bon Ange (Louis Martinié's New Orleans Voodoo Tarot Deck)
- Art (Aleister Crowley's Thot Deck)

XV/ Ego

The Ego (also represented as the Devil) represents the consequences of being too absorbed into your bliss. In other words, it represents the consequences of being absorbed in superficial or shallow desires. In the Lover card, we met characters who are in a dilemma of choosing either Sustainability or Excitement. The energy of this character is chaotic and the consequences of trying to ignore something that is beyond your control. This also represents a trait within ourselves that we may be repressing. Sometimes situations make us feel obligated to go against our nature but in repressing our true nature, we are attracting chaotic energy to surround us. This character is a reminder that we need to be true to ourselves and to ride through the chaotic events until the energy resides as chaos results in change which can be both good and bad.

- **Meaning**: consequences of residing in taboo, chaos, change
- **Element**: Earth
- **Energy**: Feminine
- **Advice**: be true to your truest nature, never substitute your identity only to "try" to fit in. By ignoring your true nature, you are leading yourself towards a chaotic future.

Tarot Deck Equivalents:

- Courir Le Mardi Gras (Louis Martinié's New Orleans Voodoo Tarot Deck)
- Art (Aleister Crowley's Thot Deck)

XVI/The Tower

A Sacred Tree is a gateway between realms; from the Sky (Cosmos) to the ground (Earth) to the roots (Underworld). The lightning bolts symbolize the "fall" of the tower, as the beings of other realms have intervened in the human world. Their intervention has caused chaos beyond our control. Signs of this can be found in the disorder madness within society. At this time, we may feel helpless. In this helpless state, we present offerings to the "gods" to ease our suffering and to return the balance. The solution to this impossible situation is surprising that the person seeking "help", holds the key that will create a change for the better.

- **Meaning**: Gateway to different realms, Autumn sacred tree, chaos, destruction, falling leaves
- **Element**: Fire
- **Energy**: Masculine
- **Advice**: sometimes the "impossible" has a solution that only you can find. The only way you can uncover this is by being true to yourself.

Tarot Deck Equivalents:

- Deluge (Louis Martinié's New Orleans Voodoo Tarot Deck)
- The Tower (Aleister Crowley's Thot Deck)

XVII/The Star

The Star is about a character who needs to reconnect with the Divine. Earlier in our development as a Hierophant, we returned to these institutions to be able to observe them from a new perspective. But this is a clear reminder that the character has dwelt too long and has become re-intertwined with these energies that they previously learned to break away from. To reach the highest level of consciousness, we must understand that it is all an illusion. This character is a reminder to let go and reconnect yourself with higher understanding and awareness and to not make the events of this reality become YOUR reality.

- **Meaning**: Hope, calm, peace, tranquility, transition of realities, detachment of this reality
- **Element**: Air
- **Energy**: Masculine
- **Advice**: cut all strings that bind you to this reality, so that you can be liberated.

Tarot Deck Equivalents:

- Z'Etoile (Louis Martinié's New Orleans Voodoo Tarot Deck)
- The Star (Aleister Crowley's Thot Deck)

XVIII/The Moon

The Moon is very similar to the Empress. They both share the same energy of nurturing and imagination and are all products of the Divine and the Divine consists of collective consciousness from all beings residing in its imagination. Each of us holds the power to tap into this Divine chain of consciousness, we can manifest whatever we want that is birthed from our imagination. In this journey of development, we have lost much of ourselves while trying to understand this reality. The Moon represents allowing your consciousness to flow into the Divine and embrace the power of imagination. It is through imagination where we can discover new paths to take to obtain ascension and prosperity.

- **Meaning**: Emotions, Imagination, Dreams, Divinity
- **Element**: Water
- **Energy**: Feminine
- **Advice**: allow your consciousness to flow through the energy of the Divine, that way you can liberate yourself and find solutions to the problems you may be facing.

Tarot Deck Equivalents:

- Magick Mirror (Louis Martinié's New Orleans Voodoo Tarot Deck)
- The Moon (Aleister Crowley's Thot Deck)

XIX/The Sun

The next stage in this journey is the Sun (Ra), this image represents you turning your metaphysical attention back to yourself. The character has gone as far as they felt they needed to go when embracing the Divine energy, now the character reverts its attention to itself. This is like the High Priestess stage, except that in this stage you are now aware that there are no boundaries and that time is nonexistent. Your character is now more aware of what to allow and what to remove in your life. Your level of consciousness is so immense, that the physical act of cleansing is useless, as your awareness puts everything in place.

- **Meaning**: Success, Happiness, Awareness of Self
- **Element**: Fire
- **Energy**: Masculine
- **Advice**: you can move mountains without even lifting a finger, as your mind now understands the concept of life without any boundaries.

Tarot Deck Equivalents:

- Gros Bon Ange (Louis Martinié's New Orleans Voodoo Tarot Deck)
- The Sun (Aleister Crowley's Thot Deck)

XX/Judgement

Our ancestors believed that to pass to the next phase after death, your heart was weighed against a feather. If your heart balanced against the weight of the feather, you were free to pass to reach the next level of consciousness (or resurrected). If your heart out-weighed the feather, then you failed the final judgment. This stage represents the removal of everything that causes you to have a heavy heart. Buddhism teaches that negativity blocks our true potential to reach the highest consciousness. This stage of Judgement represents the need to cleanse the remaining negative energy from your heart: Fear, Guilt, Shame, Grief, Lies, Illusion and Earthly Attachment.

- **Meaning**: Rebirth, new phase, intuition, balance
- **Element**: Water
- **Energy**: Feminine
- **Advice:** these feelings and energies that are weighing you down may seem important, but they are not. Learn to let them go so that you can blossom.

Tarot Deck Equivalents:

- Ancestors (Louis Martinié's New Orleans Voodoo Tarot Deck)
- Aeon (Aleister Crowley's Thot Deck)

XXI/The World

The World is the final card in the Major Arcana, but it is not the final card in the story. This stage in development marks that you have reached a higher level of self and in this new level of self, the story will start all over again. You have mastered balance with this new self and gained a new level of wisdom. The four elements are in their place and you praise your new level of understanding. But just as you gain this high level of insight, just around the corner you will become the stumbling Fool in this new stage of awareness. The story of spiritual development is a continuous cycle: from a Fool to the World.

And back again, until you go as high as you can go.

- **Meaning**: Completion, wholeness
- **Element**: all four elements
- **Energy**: Duality
- **Advice**: celebrate this stage in your life. Every moment of cosmic clarity should always be celebrated and used as a guide when you must return to your journey again.

Tarot Deck Equivalents:

- Carnival (Louis Martinié's New Orleans Voodoo Tarot Deck)
- The Universe (Aleister Crowley's Thot Deck)

In comparison is that these two decks have a slight alteration in how they are structured. Crowley's deck, for example, changes the Minor Arcana Pages into Princesses. While Martinié's deck has an additional wild card (Les Barons) that is made specifically for voodoo card reading. Furthermore, the titles in the Major Arcana cards were also changed in these comparative decks.

Regardless of these changes of titles, the Energy surrounding the cards is the same. It is good practice to look at the card decks comparatively, to get a better understanding of the function of the cards. Some decks will have more symbolism associated with it, like Crowley's. While other decks will focus more on depicting the overall feeling of the card, like Martinié's.

After the sections that explain the cards, there are spreads that you can use to perform your reading. This is not the limit of what you can do with your tarot cards. Beyond the readings, I would highly encourage you to set up your Astro Natal Chart, in which you see the planetary placements at the time you were born. Once you create your natal chart, you can begin placing the tarot cards that are in correlation with the zodiac symbols on your natal chart. This will help you further understand your characteristics and your story. It will also help you understand how the cards work with each other.

Minor Arcana

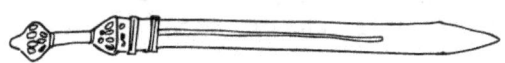

House of Swords – Elder God of Air

The **House of Swords** is a suit in the Minor Arcana. On the Tree of Life, this element represents *Yetzirah* (in Kabbalah), **Pet** (in Kemetic) or **Air**. This level of awareness is one step above our physical awareness, Earth, but not high enough to be Spiritual ZPE. The realm of Air plays a major role in how we perceive the world around us.

Kings of Swords
The characteristics of this card represents masculine energy associated with a character that passes judgement on to others. This character can also be a mentor and a mediator of conflict. This character is very reserved and does not easily showcase his emotion nor is he easily moved by others who are exceptionally emotional. He is very even tempered, observant and speaks last. Because of these traits, he is often deemed to be very wise.

Advice: Listen to the inner wisdom offered by the wise elder that dwells inside of you.

- **Positive Characteristics**: Rationality, wisdom, judgement, resolution, will, science, critical thought, logic and intellect.
- **Negative Characteristics**: Unaffectionate, vagueness, and detached.

Queen of Swords

The characteristics of this card represents feminine energy associated with a longing wife, who has high standards towards life. Her high standards do not make her hard to please, but rather, more specific about the direction she wants to go. With that said, her true love lies within the desire to refine the world and to open the 3rd eye of her people. She desires that everyone finds balance and peace and space so that they can grow. The Queen of Swords is very intelligent, independent and powerful. These characteristics amplifies her sensuality, making her very desirable even if she is not one to be easily controlled.

Advice: Be confident in making your own decisions and being independent. Do not fall victim to the false ideology that you must conform to an institution. Be true to yourself and you will attract those that are deserving of your presence.

- **Positive Characteristics**: Wisdom, independence, free thinker, vast imagination, confidence, intuition and sincerity.

- **Negative Characteristics**: Apathetic, sadness, privation, solitude, deceit and vagueness.

Knight of Swords

The nature of this card is associated with the mindset of a restless and aggressive character, who has a hostile attitude towards everything and everyone. This character is constantly pointing fingers at everyone else but himself, as he considers himself to be righteous. This character doesn't look to resolve problems with negotiation, but rather seeks to attack any and everything, then ask questions about it later. Because of this mindset, the Knight of Swords will often do the right thing, but the true motive behind the actions is destructive.

Advice: Be accountable of your actions and seek to ask questions before leaping into action. It is okay to take a break to contemplate about things before you lunge forth.

- **Positive Characteristics**: Quickness, heroism, creativity, steadfast, courage, nimble, visionary and determined.

- **Negative Characteristics**: Sweet talker, unrealistic, harsh, unreliable, harsh and deceitful.

Page of Swords

This character goes by many names, such as Princess, Watchman and the Spy. The energy surrounding this card is feminine. She represents diplomatic negotiations between two powers. Though her position appears to be lowly, she is very powerful as she holds the power to decide how the information she has gathered will be used. Should she choose to, she could use the information to hold other powers accountable or for personal gain.

Advice: There are situations that require you to remain anonymous. Even in these situations you hold the key to decide if you choose to remain neutral or if you choose to make a move more in line with your personal ambitions.

- **Positive Characteristics**: Curious, cautious, brave, professional, serious and trustworthy.
- **Negative Characteristics**: Potentially spiteful, gaiety, unaware of potential, wasteful, and ignorant.

Ace of Swords

This card is the mastery of the element of this suit, Air. The characteristic associated with this card is mastering the ability of knowing the difference between reality and fantasy (or Knowing verses Believing). In the BLK Excellence tarot deck, this card is depicted by a Kemetic version of King Arthur, reaching for the sword Excalibur. This depiction represents a character coming face to face with their potential and being true to themselves. In this leap of faith, we have to make sacrifices which is usually just the acknowledgement that things are destined to change. This is the energy surrounding this card.

Advice: Remain focused and true to yourself. You are about to make a major move closer to your desired destiny. Do not faulter, the end result will be worth it.

- **Positive Characteristics**: Becoming in sync with black energy, intuition, realization, intellect and destiny.
- **Negative Characteristics**: uncertainty, doubt, easily redirected and distracted.

Two of Swords

The characteristic of this card is associated with a clash between ideas or ambitions. Similar to the chariot, in which these ideas will take you different directions, forcing you to be stagnant. The difference in this element is that these ideas are trying to communicate or meld together but are failing at these attempts to make these ideas work together or coincide harmoniously.

Advice: Evaluate all of the facts and make sure all of the information received is lined up before acting or deciding. In situations like this, it is often to better to be patient and wait for the right time to make a move.

- **Positive Characteristics**: Peace, friendship, alliance, stability, coordination, balance, observant and visionary.
- **Negative Characteristics:** Indecisive, crossroads, stagnant and confusion.

Three of Swords

This card represents the separation of a significant connection or relationship. On one aspect this could be a separation from a connection that could have been destructive to your own wellbeing. As a result of this separation, the character isolates themselves in order to find themselves. The character shows detachment from former ties and goes out of their way to find new beginnings.

Advice: In some cases, detachment is necessary so that we are able to heal. If there is a relationship (work, love, friends, etc.) that seems destructive, then the best move would be to detangle yourself from the roots and find a new perspective so that you may start again.
It might be healthier to disentangle yourself and start fresh.

- **Positive Characteristics**: Realization, new beginnings, intuition, ambition, contemplation of ambition and change.
- **Negative Characteristics**: Disappointment, distance, pain, separation, withdraw, melancholy, isolation and uncertainty.

Four of Swords

The characteristics of this card represents retreating to a place of serenity. In this place, you are surrounded by the energy of your ancestors and/or guardians and are comfortable with evaluating your existence. This card represents the journey to a sacred place in hopes of finding an answer, like the ancient temples in China. High up in the clouds, you are surrounded by ancient energy that's only interest is to help you succeed in your quest.

Advice: When you are becoming anxious, chose to reconnect to your Eternal Self and train yourself to mentally detach yourself with the on goings of the physical realm. The opinions and prejudices of others should not concern you as their opinions only belong in the physical realm, not the spiritual realm. As a result of their opinions only existing on Earth, you should take note that those opinions are temporary and bound to dissolve into nothingness.
- **Positive Characteristics**: Manifestation, internal reflection, serenity, spiritual guidance, tolerance and intuition.
- **Negative Characteristics**: Isolation, exile, misunderstanding, confliction, self-consciousness and doubt.

Five of Swords

The characteristic of this card is similar energy of that of a great loss, as if a loss after a battle. The war has neither been won or lost, however, the grief associated with the great number of causalities, linger in the energy of this card. In this great loss, the one who lost the battle has revisited the battlefield. The character returns to the battlefield to reevaluate what happened. Standing beside this character is an ancestor, who is guiding the character to see every clue, every weak spot, every blind spot and so on. The characteristic of this card may not necessary be only a representation of war, it also represents every battle we have ever fought, from work, love and other personal struggles.

Advice: Use the knowledge that you have learned from this loss to not only empower your life choices but to also empower others. Strive to trigger the highest level of intellect and awareness in others. By doing this you are ensuring that there will be less battle losses and more battle wins.

- **Positive Characteristics**: Analysis, rational, philosophical, mentor, observant, optimistic and decisive.
- **Negative Characteristics:** Defeat, betrayal, fear, grief, loss, disappointment and overly cautious.

Six of Swords

The characteristic associated with this card is in harmony with the *Tehuti Principle*, in which the character of this card is able to research superstitions and turn them into scientific facts. It can also represent the process of transferring theories into laws and credible concepts. In other words, the act of manifesting cosmic thought. In the process of doing so, the character of this card becomes a Navigator to others who are also seeking a foundation to build upon. As a Navigator the character is able to travel through the sea of black energy and knowledge to arrive at their desired destination. The character holds a special or unique or a different type of wisdom that many will fail to fully understand but will rely on in moments of need.

Advice: Do not falter your journey because those around you may not perceive the bigger picture. You are not weird or strange. You simply have a unique ability to look at the world in a powerful way. You see how things work and come together, which most may not immediately understand. However, the results of your actions will become the foundation others will need. Your action will rescue many who are drowning in fantasy, dogma and superstitions. Stay true to your path and trust in your judgement.

- **Positive Characteristics**: Deep critical thinking, travel, analysis, science, success, determination and rationality.
- **Negative Characteristics**: Anxiety, lonely, difficult road, distrust, low self-esteem and doubt.

Seven of Swords

The characteristics of this card is around the energy of optimism, in which the character of this card is lost in a dream or rehearsing his ideas regardless of the consequences. The character is very sure of himself that there is no thought that he may lose. Typically, the image depicts a character who has snuck into an enemy camp site at night. In this setting, the main character not only sneaks into the site but also manages to steal the enemy's swords... disarming the enemy before the enemy decides to strike. In the BLK Excellence deck, the main character is Esu, which is an Orisha. He has entered into the campsite and stole the swords of the supposed heroes. The character Esu is a trickster but he also holds the keys at the crossroads. Which means that he holds the power to determining who can make a move.

Advice: Forward think through all challenges. Try to look at things ten steps ahead, so that you can control the outcome. In the event that you are forced to make a move that will empower you, do it. Some risks are worth taking so long as they produce constructive results.

- **Positive Characteristics:** Cleverness, hope, astuteness, forward thinking, observation, unmasking lies and confidence.
- **Negative Characteristics**: Trickery, bad faith, deceit, setting traps and mischievous.

Eight of Swords

The characteristics of this card surrounds the energy of a *Test* (which is the nickname for this card), where the character is subjected to harsh judgement to whether or not they are worthy. This characteristic echoes experiences in our life where we are put under close inspection by others. The character ultimately will either pass or fail.

Advice: Do not let the surveying of others determine the value that you see in yourself. Rise to the challenge and be confident in your abilities. Regardless if you pass or fail, the experience alone will strengthen you.

- **Positive Characteristics**: Courage, determination, focus, excel, success and acceptance.
- **Negative Characteristics**: Sensitivity, blame, difficulty, falseness, solitude, news, uncertainty, self-conscious, doubt and impressionable.

Nine of Swords

The characteristics associated with this card, surrounds a character who is paying homage to a significant loss. Those that the character is mourning, left the character alone to go seek out glory. When their act was done, this left the character in shambles and vulnerable to the world. The character represents the victims who suffer from egotistical actions done by others. In other words, the price of Pride.

Advice: There are situations where we will devote all of our time and resources for projects or actions that just won't work. We are driven to keep sacrificing more of ourselves for the sake of pride. Or we are the victims of others pride. Be honest with yourself and others, when a situation clearly doesn't work. It is better to walk away from something damaging than remaining in a situation for the sake of personal pride.

- **Positive Characteristics:** Awareness, analyze, acceptance and forward thinking.
- **Negative Characteristics**: Pride, apprehension, mourning, crisis, pain, blind faith, self-destructive and arrogance.

Ten of Swords

The characteristics of this card is associated with the energy surrounding a tremendous finale. There is no revival in this situation, as the enemy has been defeated or the line has been crossed that there is no redemption from the action. This energy brings a needed closure and release from an intense situation. There no longer is a need to wait or wonder, the steps have been taken to make a change, and the change has come. With that said, in moments like this, you are free to let go. In the BLK Excellence tarot deck, this card is depicted with the freedom fighters in Mexico, who are led by Gaspar Yanga. Gaspar Yanga won his battles against the Spaniards and was able to build a free city.

Advice: The Storm has finally come. The only thing that you can do is protect yourself and those that are important to you. Be mindful that eventually the Storm will pass by and leave behind destruction that will require you to rebuild.

- **Positive Characteristics**: Transformation, rebirth, change, freedom, victory and destiny.
- **Negative Characteristics**: Ruin, grief, affliction, loss, destruction, pain, catastrophe and uncertainty.

House of Cups – Elder God of Water

The **House of Cups** (also known as the *House of Chalices*) represents the Elder God of **Water**. As we learned earlier, the characteristics of Water is connected to our emotions. This means that the way that water is perceived in the card is meant to reflect the emotional journey we have all experienced in one time or another. Our emotions are the source of our creativity and lead us to obtain wisdom. The realm of our emotions lies in **Briah** (in Kabbalah) or **Duat** (in Kemetic).

King of Cups
This card is associated with the energy of a master of emotions, holding a cup of immense power and mystery. The image depicts emotional energy bubbling out of the goblet that he is holding, like the mouth of a watery volcano. Should this character wish to, he can release immense power that will move the world around him. He holds the power to tap into the hearts of others and is sometimes seen as a shaman or a holy man. The king can decide who or what can have access to this power, as he is selective, only those he deems worthy can drink from this goblet.

Advice: Should you choose to, you can tap into true and immense power, that can move legions and reshape your surroundings. You must decide who or what is worthy to receive your wisdom.

- **Positive Characteristics**: Emotional control, generosity, equilibrium, honesty, clemency, creative literature, charity and wisdom.
- **Negative Characteristics**: Manipulative.

Queen of Cups

The characteristics associated with this card is feminine energy that portrays a powerful character who is not only in control of her emotions, but it abundantly compassionate towards the world around her. She is very empathetic which is both her gifts and her weakness. Her empathetic nature leaves her vulnerable to the emotional needs of everyone else, making it hard for her to identify what actions are in her best interests. Similar to the King of Cups, she also holds a similar goblet, that is bubbling with immense emotional power. Unlike the King of Cups, she chooses to share her power with anyone.

Advice: Being empathic of others is great, but it is also important to recognize that you also need to be aware of your needs. When supporting others, do not let their emotions overwhelm you, be supportive from a safe emotional distance.

- **Positive Characteristics**: Imagination, friend, virtue, compassion, creative, sensitivity, empathetic, leader and nurture.
- **Negative Characteristics**: Vulnerable and impressionable.

Knight of Cups

The nickname for this card is Homecoming, as it depicts a hero returning home from a long difficult journey. This hero has quenched his thirst for adventure and now is exhausted from fighting, seeking to return to a place surrounded by love and familiarity. His adventure and battles have made him older and wiser, and now he is ready to bring the knowledge he has uncovered, home. With that said, this character believes that the place he is returning to will be the same as how he left it. This is the weakness of the character. While he changed from the world, the place he is returning to has also probably changed.

Advice: Even though you are exhausted from fighting your battles (whether in searching for love, wealth, justice, etc.), returning back to where you came is a very important moment that requires all of your strength and energy to complete. Be open to the changes in your reality.

- **Positive Characteristics**: new perspective, messenger, arrival, new love, marriage proposal, wisdom, contemplation, compassion and connection.
- **Negative Characteristics**: exhaustion.

Page of Cups

The characteristics associated with this card is surrounded around a character who is naïve towards the dangers around them. The character is easily tossed about due to external energies, but the character continues to remain aloft of the gravity of the situation. The character is depicted by a young man who is dressed in finery. The page also holds a lot of power but is unaware of what is in their possession. The weakness of this character is that they need to develop critical thinking abilities.

Advice: It is okay to make yourself available for different situations, however, be aware that being devoted to these situations may put you in risky situations. Therefore, be mindful of the events you are putting yourself in as well as be aware of the power that you hold. You also hold the power to change the direction of a situation, should you understand how to use it as well as choose to use it.

- **Positive Characteristics**: Friendship, new beginning, loyalty, sympathy, empathy, and dedication.
- **Negative Characteristics**: Irrational thinking, naivety, hysteria and solitude.

Ace of Cups

The characteristics represents the infamous goblet of emotional infinite power. It represents a source of Black Energy and a connection to the heart. Having access to this power gives you the reassurance that you can relax or meld into love, support and companionship.

Advice: We are presented with meeting new people or being put into new situations that leave us to question whether or not we should be trusting. This card reminds us that there are hidden gems in people and events that we should try to find, explore and accept.
- **Positive Characteristics**: new beginnings, compassion, devotion, intuition, and open-mindedness.
- **Negative Characteristics**: doubt, uncertainty and denial.

Two of Cups

The characteristics of this card is associated with the union of two becoming one. It is depicted with two lovers becoming united, but it can also represent friendships, partnerships or a mutual understanding. This card can also represent a resolution of a personal conflict that you may have been struggling with.

Advice: Express yourself. Allow yourself to build constructive connections with others and/or yourself. Let those that are significant to you know that you care for them. By doing this, you are strengthening the bonds.
- **Positive Characteristics**: Truth to yourself, rational, confidence, compassion, trust and acceptance.
- **Negative Characteristics**: vagueness, suspicion, mistrust and anxiety.

Three of Cups
The nickname for this card is "Yes." The card typically depicts three women celebrating their sisterhood under a full moon. This represents mutual emotional support and mutual admiration.

Advice: There is a support network that you maybe overlooking. Bring this team together and witness the benefits of this unity.
- **Positive Characteristics**: Teamwork, family, unity, harmony, balance and support.
- **Negative Characteristics**: Hesitation

Four of Cups
The characteristics associated with this card, is the energy of a character who is mentally and emotionally restless. This person may be emotionally uncomfortable or dissatisfied. Due to the situation, the character is stagnant and left pondering about what steps they should take.

Advice: An inaction is still a form of action. If you find yourself surrounded by a negative emotional feeling, then it is a sign that the situation needs to be changed.
- **Positive Characteristics**: Contemplation, reflection and awareness.
- **Negative Characteristics**: emotional turmoil, indecisiveness, pain and stagnant.

Five of Cups

This card represents the after effect of an emotional uproar, like a rage or outburst. The consequences of such an event would be negative, such as abuse, broken relationships, poverty or withdraw. The cause of the outburst is potentially triggered through loss and frustration.

Advice: It is better to accept that loss is a part of physical life, but also, whatever we lose it is reborn in a different state. By understanding this, there will no longer be a need to be frustrated or at a loss that may lead you to be self-destructive.

- **Positive Characteristics**: Awareness and recovery.
- **Negative Characteristics**: self-destructive, aftermath, chaos, frustration and outburst.

Six of Cups

The nickname for this card is *The Past*, as it represents a character who is reflecting on our nature and identity. The essence behind this card is for the character to tap into the person they were before, back when they were optimistic and had a fresh perspective about life.

Advice: In life we are forever changing, which is inevitable. But sometimes we change so much that we have distanced ourselves from our true nature. When hardships and struggles come our way, we tend to become skeptical about life. This card reminds us to look back to our past characteristics before we were tainted from anguish, and to use that youthful mindset as a guideline to manage our present circumstances.

- **Positive Characteristics**: curious, contemplate, thoughtful, self-evaluation and a scholar.
- **Negative Characteristics**: stagnant and dissatisfaction.

Seven of Cups

This card is nicknamed *Fantasy.* It was given that name because it depicts the act of envisioning an alternate future or circumstance. The future is not set in stone, the actions that the character does today will affect the results of tomorrow. Therefore, this card encourages the character to dream of different possibilities in hopes to finding a new direction.

Advice: There is power behind dreams. As we learned in this book, it is in the realm of Water where we can conspire amazing wonders and possibilities. Allow yourself to daydream and perhaps set aside time each day to do just that. After daydreaming, jot down the ideas and concepts that you imagined and see if those ideas will resolve the circumstances that you currently face.

- **Positive Characteristics**: Dream, fantasy, black energy, unbiased and tolerant.
- **Negative Characteristics**: limited amount of time and external influences.

Eight of Cups

The characteristic of this card is associated with the energy of a victim who has just been molested or attacked. After experiencing this trauma, the woman is cast aside by her attacker and left alone to wander through her town. As she wanders the streets, she is left vulnerable and in shock from the experience. This energy represents the aftermath of a sudden emotional shift that has left us distraught. Though the character is seemingly surrounded with vulnerability, she also holds a strength which is her determination to make it through or to make it home.

Advice: There are circumstances that are not necessarily as extreme like molestation or being attacked but bare a similar reaction of being distraught and emotionally stunned. In most of these cases, the event is not your fault. In these circumstances you have a choice to give up and remain vulnerable or continue on your path to safety. Hold on to your determination to find a haven where you are safe to regather your emotions and thoughts. Allow yourself all the time that you need, in which to come to terms with the experience. Once you finally come to terms, you will be ready to begin formulating a solution for yourself.

- **Positive Characteristics**: Determination and hope.
- **Negative Characteristics**: Pain, anguish, fear, uncertainty, lack of trust in others and/or environment.

Nine of Cups

This card is nicknamed *Victory*. The characteristics of this card is associated with the energy of the feelings of achievement. It refers to a character who has worked diligently and has finally reached a point to which they have finally found accomplishment in their work. In this moment, the character feels a sense of gratitude, happiness and of course, victory! This is but a temporary feel that has been well earned.

Advice: Allow yourself time to relax and enjoy this moment of accomplishment.
- **Positive Characteristics**: victory, content, grateful, peace and balance.
- **Negative Characteristics**: temporary

Ten of Cups

The characteristics associated with this card is the energy of a communion of a family, in which many generations come together in harmony (as depicted by the rainbow – signifying the end of hard times).

Advice: Be open to connect and share with all generations of people.
- **Positive Characteristics**: Pregnancy, adventure, love, support, unity, abundance and balance.
- **Negative Characteristics**: Struggle with commitment

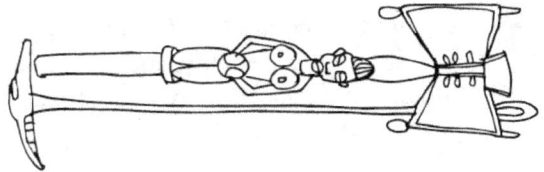

House of Wands – Elder God of Fire

The **House of Wands** is a suit in the Minor Arcana. The wand dwells in the realm of Fire, which represents the events of the spirit and soul. Each card in this suit represents a stage or an experience that we encounter, spiritually. In the Tree of Life, this suit represents the realm of **Atziluth** (in Kabbalah) and the source of wisdom, **Chiah** (in Kabbalah) or **Nunu** (in Kemetic).

King of Wands
The characteristics of this card is associated with a charismatic king, who is dynamic as well as eager to leave an impression on the world. His determination to expand his ambition for adventures and taking risks. He easily gets bored of the routine and is constantly seeking a change of pace. He is not a follower, by any means. His Ego keeps him in the lead as a sympathetic ruler.

Advice: there are no specific obstacles preventing you from reaching your goal. Be mindful of the aftermath of your decisions.

- **Positive Characteristics**: Confident, sympathy, focus, fidelity, security, innovation, adventure and charisma.
- **Negative Characteristics**: pride, ego and arrogance.

Queen of Wands

She has a natural ability to delegate and balance different situations, because of this skill, this character is an inspiration to others. She works behind the scenes (the spiritual divine scenes) to ensure that all the "T's" are crossed and "I's" are dotted, which makes her very reliable. Or in other words, she spiritually guides us to pay attention to the necessary details that we will need in order to become successful. She is a very humble and playful character, who doesn't mind rolling up her sleeves and getting into work with her workers. She is devoted to trying to save or change the world from the spiritual realm.

Advice: You may not be in a leadership position, as you would like. But you should allow yourself to get more involved with supportive tasks that help to keep on target.

- **Positive Characteristics**: Charisma, modesty, multi-tasking, nurturing, supportive, humble, fertility and balance.
- **Negative Characteristics**: introverted and self-centered.

Knight of Wands

The characteristics associated with this card is the energy of a short-tempered Knight. This character has a chip on their shoulder, and is looking to start a fight, which is very similar to the Knight of Swords. In fact, both characters would ideally be the best of friends. The difference is that the Knight of Wands is more of an Absolutionist, while the Knight of Swords is just seeking excitement. The Knight of Wands believes that they are righteous and any actions that are done, is helping to save the world from itself. The Knight of Wands is, in some case, a religious fanatic or crusader. His believes his will is adorned and that he is the savior of the world. Many may see him as a hero, but others are fearful to be caught in his path.

Advice: We are surrounded by so much knowledge that it is easy to feel that you are right, and others are wrong. By seeking the world in this way, your actions may be destructive to others, even if you remain unaware of the consequences. Allow yourself to be humble and learn from others, regardless of how hard it may be to listen.

- **Positive Characteristics**: Travel, spirited, courage, powerful, victory, leader and reformation.
- **Negative Characteristics**: narrow-mindedness, impatience and anxiety.

Page of Wands

The characteristics of this card is the energy of a unique individual. This character requires little affirmation from others. In fact, the energy of this character is that they reserve their power and resources for themselves until others begin to serve them. This character is very independent and their freedom from institutions is essential to their spiritual wellbeing. They would rather starve than be forced into a servitude that comes with a great cost to their freedom. The Page of Wands may seem humble, but in reality, this character is a potential ruler and leader. Their future is uncertain as it depends on their mood of the situation. For this reason, the energy associated with this card is like a wild card.

Advice: While you go about your daily tasks, you are exposed to information that will help you advance later on. With that said, carefully survey these situations for useful information that will later assist you to abundance. This information could come in many forms, such as an internship, mentoring, apprenticeship, working under a more tenured professional, so that you can learn the tricks of the trade, and use those skills to advance your life forward.

- **Positive Characteristics**: Observant, intellect, analysis, confident, sincerity, news, surety and focused.
- **Negative Characteristics**: Arrogant and overly critical.

Ace of Wands

The characteristics associated with this card is the energy of a catalyst. But not the form of catalyst that will create havoc in your life, but rather, a series of events that create a change or shift towards your goals. Change is a part of a birth of new awareness. In light that this card is associated with Fire and the Spirit, this change will be more intuitive and divine. In other words, a shift in perception. This shift will lead you to be able to resolve issues, complete projects or build a foundation that supports your ideals.

Advice: Be ready for a shift in awareness and remain open to inevitable changes.
- **Positive Characteristics**: Change, birth, new beginnings, origin, resolution and instinct.
- **Negative Characteristics**: uncertainty

Two of Wands

The characteristics associated with this card is surrounding a spiritual divide in which you are divided between a pessimistic or optimistic mentality. In other words, the mindset associated with you feeling that you can neither win nor lose. You are feeling as though you may be at your limits or at a standoff with another person. There is doubt that lingers over the energy of this cards, in which you may be feeling as though you have gotten way over your head, in a situation.

Advice: In this situation, it is okay to take a step back to breathe and be observant about the problem at hand. Some decisions do not come immediately . In these cases, you need to be patient for the answer to reveal itself.
- **Positive Characteristics**: Self-awareness, reflection, analyze, evaluation, crossroad and patience.
- **Negative Characteristics**: stagnant, indecision, controversy, doubt and overwhelmed.

Three of Wands

The number three symbolizes creation and/or new beginnings. In relation to the Suit of Wands, this new creation can only be associated with the spirit. The energy characteristics associated with this card are balance and an eagerness for new endeavors. The character has, in a way, finally come to the right moment to make a move and is giving thanks for coming to this understanding.

Advice: Trust your intuition and make a move.
- **Positive Characteristics**: Optimism, achievement, balance and success.
- **Negative Characteristics**: self-centered

Four of Wands

Similar to the Three of Cups, this card symbolizes a spiritual union of a team. The energy associated with this card is that a few energies are coming together to build something sound. Since the Suit of Wands is associated with the spirit, this means that this union is building a spiritual-driven foundation that could potentially span over time and space, or in other words... it will be enduring. This could be a business, a community project, a creative program or even an educational assignment. In either setting, it is a collaboration of leaders who are coming together to build together.

Advice: In this setting avoid becoming the dominant one. This is a situation where it is a collaboration between leaders, where everyone's

voice needs to be heard so that this project can be successful.
- **Positive Characteristics**: teamwork, business, building, busy, dynamic, innovative confidence, marriage, and union.
- **Negative Characteristics**: misguided

Five of Wands

The energy surrounding this card is the energy of a character who, at the expense of others and themselves, are so ambitious for pushing their agenda. In our push to strive for personal excellence, we may undoubtedly make others the losers or… if we fail to meet our high expectation, we will in turn label ourselves as the loser. In the BLK Excellence deck, this card is depicted with the spiritual energy of Shango, a Yoruba deity. This deity was incredibly powerful and determined to maintain dominion over the village called Oyo. In his fight to win, he was not mindful of the harm he was doing to himself or to others. When he lost, he fell to so much grief that he hung himself from a tree. The Yoruba have spiritual dances that invoke the energy of these spirits. In the dance of Shango, his moves are powerful and with so much charisma. But in every step of his moves, he is blind to the damage he could be potentially doing to himself and those around him. This is the energy of this card.

Advice: When you are fighting so hard to produce trivial results, who is benefiting? Are the results that you are seeking beneficial or destructive to your reality and those who are in it? Take time to visualize your goal and to brainstorm the pros and cons of the situation. If the situation has more cons than pros, then restructure the goal to where the benefit becomes a mutual exchange for your hard work.

- **Positive Characteristics**: Determination, comfort, power, ambition, drive, focus, and strength.
- **Negative Characteristics**: self-destructive, fragility, hot/cold mindset, struggle

Six of Wands

The energy surrounding this card is the energy of the "Sacrificial Lamb." The character has devoted themselves to resolving the problems within their community. In most cards, the character is depicted going through a victory parade as the hero of a great battle. In the BLK Excellence deck, the card depicts Kimpa Vita burning at the stake in the medieval Congo. Kimpa is a historical freedom fighter of the 17th Century. During a time when the Congolese people were being kidnapped by colonialists and shipped out to the Americas as slaves, the Congo was left in shambles as the land was stripped of the native people. Due to the kidnappings, the residences left their capital city and lived in harsh conditions in the wilderness. Kimpa devoted her life to peacefully bring the people back to their beloved capital. Thousands of her people marched peacefully with her. Kimpa was later captured and charged with blasphemy by the Portuguese cleric, then burned at the stake.

Advice: Whether at the end of your battles you are able to witness the victory of your accomplishments, know that in your heart you achieved your ambition. Your desire to fearlessly

act on circumstances that spiritually move you, will inevitably inspire others to support your dreams.
- **Positive Characteristics**: pious, justice, empowerment, focus, quest, merit, victory, community and compassion.
- **Negative Characteristics**: self-sacrifice

Seven of Wands

The energy surrounding this card is the energy of fearless leaders who are almost always successful with moving their will throughout the world. This character is memorable to others, as they spiritual stand above the rest. They have no boundaries as they create new limits that exceed the expectations of others. In the BLK Excellence deck, the image depicts freedom fights in the Caribbean and the Americas (Tacky, Queen Nanny, Queen Cubah, Nat Turner and Harriet Tubman). These heroes were able to change the expectations of those around them, so much so that the system was forced to change because of the results of their actions.

Advice: Do not place limits on your competitive nature. Allow your competitiveness to excel and push you beyond your own expectations.
- **Positive Characteristics**: Willpower, competitive, focus, results, innovative and change.
- **Negative Characteristics**: Ego and obstacles.

Eight of Wands

A swift unfolding of events is about to be revealed. These events could be either out of your control or intentional results. The energy of this card is depicted by arrows being launched at the character from an unknown source.

Advice: Prepare yourself for whatever is quickly approaching you.
- **Positive Characteristics**: Rapid response, quick thinking, preparation, good intentions.
- **Negative Characteristics**: Distraction

Nine of Wands

Spiritually exhausted, the character is healing from the wounds they have taken. In this moment of waiting to heal, the character is reflecting on their victories. The characteristics of this energy suggest that you pass the torch on to your comrades, friends, family or colleagues; so that they make continue the fight while you are recuperating.

Advice: Even our spirit needs time to heal from all of the battles it has been fighting. Step away from the leadership role and delegate your work to others. Once you are healed, you will be ready to fight again… harder, faster and stronger.
- **Positive Characteristics**: Wait and patience.
- **Negative Characteristics**: Delay, defense and aggression.

Ten of Wands

The character is committed to a task which demands all of their focus and energy and resources, once false move will leave you in a vulnerable situation. The character of this card is busy working and building up the walls of his home. He is using all of his resources to make his home secure from robbers.

Advice: The work that you are doing is very delicate and requires your determination to see it through. Maintain a fresh state of mind, no matter how vigorous the work is becoming. Once it is finished you will have your vision secured.

- **Positive Characteristics**: Focus, determination, busy
- **Negative Characteristics**: Concern for the unknown, bitterness, bad faith, deceit.

House of Coins – Elder God of Earth

The **House of Coins** (also known as the *House of Disks* or *Pentacles*) is a suit in the Minor Arcana. The symbol of coins represents the Elder God of **Earth**. How the coins are presented in the card represents the various events we experience in our physical perception of the world. In the Tree of Life, this house represents the realm of **Assiah** (in Kabbalah) or **Ka** (in Kemetic). This is the "physical plane" of our reality, which can also be identified with the Root Chakra (Geb) and the Earth Chakra (Aye). Everything associated with this suit, are material or something physical.

King of Coins
The energy associated with this card depicts a King who has mastered the skill of formulating wealth. He has accomplished a lot of things in life, and he is in the position to show off his decorated accomplishments. His genius is that he not only understood the ways of the world, but he mastered materializing his ambitions and visions. He is a builder, who has excelled in the Tehuti practice of converting ideas from the cosmos into

this reality. At this level of material power and intellect, he has the power to mold this reality. Civilizations can either rise or fall by his order. In the BLK Excellence deck, this card is depicted by the historical character named King Mansa Musa I. King Mansa was (and still remains) recorded as the wealthiest man in human history. He was so rich, that he gave away his money throughout Europe and still remained the richest man. His ability to master coin gave him unspeakable power which could determine the rise and fall of civilizations, of which it did. Civilizations were formed off his charity, as well as civilizations were crushed also because of his charity. His wealth spanned over space and time and inevitably made the continent a target to invaders.

Advice: "*Fake it until you make it*" is a golden rule of life. This rule does not make you a liar. This rule forces you to learn how to access your imagination and to start to play "make believe." By doing so, your confidence will increase and be projected onto others, which in turn will help you convert dreams into materialized foundations.
- **Positive Characteristics**: Business, power, Tehuti Practice, dynamic, influential, master and result.
- **Negative Characteristics**: no control over who will be drawn to you.

Queen of Coins
The characteristics associated with this card is a Queen who has mastered critical thinking and problem-solving. She is a natural mentor and Life Coach, who seeks to enrich and empower people with the knowledge that she has collected. She is

traditionally depicted as an Oracle or a Seer, as her unique ability to calculate the probable results has either made her appear to be a psychic or has caused her to become in tuned with her psychic abilities. Either way, she has a depth of understanding of how things works but primary she is skilled with understanding how situations can be fixed. She is naturally involved with community outreach activities, like charity, healing, teaching and so on. Similar to the other Suit Queens who all desire to save the world in their reality, she desires to share her skills as she doesn't feel that her abilities solely belong to her. Whereas the Queen of Cups would supply the world with emotional energy and the Queen of Wands would enhance the world's spiritual intuition… The Queen of Coins desires to save the world through tangible means.

Advice: Strive to build a better world and in turn, this better world will reform you to be a better version of yourself. There are skills that you have that the world can benefit from. It may be the case that you are holding yourself back from being true to your nature. Be confident with being true to yourself and with speaking the truth.
- **Positive Characteristics**: Psychic, compassion, savior, healer, certainty, fertility, charity and alliance.
- **Negative Characteristics**: Denial.

Knight of Coins
The energy surrounding this card is of a character who is a farmer and/or gardener. He is a retired fighter who has put his armor and weapons on

the shelf. At any given time, he may be called back to war but in those times in between battles, he is tending to his farm and gardens; cultivating new life. Even his supposed War Horse, is more of a Plow Horse that is built to move steady and slow. This old knight has found the highest quality of life and has become relaxed, taking his time to savor every moment.

Advice: Quality moments can be found outside of battles. It is true that we may find glory in winning wars, but we find satisfaction in savoring the moments and enjoying life. In between your battles find the time to slow down and reflect.
- **Positive Characteristics**: Stability, luck, determination, balance, monk, wisdom, patience, guide and caretaker.
- **Negative Characteristics**: temporary tranquility

Page of Coins

The energy associated with this card is similar to the Queen of Coins, in which the Page is fascinated with solving problems. Unlike the Queen of Coins, the Page of Coins is focused on learning the properties and laws of nature. She is a natural scientist, who is head down and continuously cultivating and experimenting the results of solutions. She wants to understand the underlying function of success and abundance. Whether she is wealthy or poor, this character is actively building a game plan to secure her victory… but before victory can be achieved, she just has to keep working and keep collecting the knowledge and experience she needs to produce results.

Advice: Remain open-minded as you test out all of your theories and recorded answers. Whether your results are right or wrong, the action of solving the problem is developing you to become a master of this skill.
- **Positive Characteristics**: Aspirations, science, experimentation, results, apprentice, student, determination and problem-solving.
- **Negative Characteristics**: Solitude.

Ace of Coins

The energy of this card surrounds the energy of a seed that has been planted in a garden. This gift can potentially grow into something amazing, so long as it is handled with care and tended. What this seed could potentially grow into, will help you advance in your future. This is the first step towards a new goal that will become the centerpiece of your accomplishments. Since the suit of coins refers to tangible experiences and things, this "gift" could be a talent or a plan that you should begin to develop.

Advice: With the long-term goal in mind, begin to take the first steps. These steps may start off small but in the long run they will become major strides in your life.
- **Positive Characteristics**: Patience, baby steps, time, long-term planning and fortune.
- **Negative Characteristics**: no clear direction.

Two of Coins

The playful energy surrounding this card is memorizing but is destined to produce unfortunate consequences. The two coins are depicted as one being Heads and the other being Tails, symbolizing the Yin and Yang energy tied to this characteristic. The characters in this card are playful and are sometimes depicted as a juggler. As fun as it may seem a single misstep will cause a shift in the situation.

Advice: There is a time to play and a time to be wise. Avoid falling into the trap of premature actions and choices.

- **Positive Characteristics**: News, harmony and charisma.
- **Negative Characteristics**: Distractions, immaturity, goofy and fun.

Three of Coins

The three symbolizes creation or the reproduction of an even balance of masculine and feminine energy. The Suit of Coins represents the tangible elements that we associate with. Therefore, the Three of Coins represents the product of a balanced (or mastered) craft/skill. The nickname for this card is Genius, and it is typically depicted with a master craftsman leading the production of a beautiful wonder. The wonder is mostly depicted by a stained-glass window.

Advice: Embrace the eagerness of showing the world your talent and skills.
- **Positive Characteristics**: Perfection, confidence, skill, overcoming trials and nobility.
- **Negative Characteristics**: self-centered

Four of Coins

The characteristics associated with this card surrounds the energy of the two-edge sword paradox. The characters in this card have been well supported by others to the point of reaching material prosperity. It is the results of reaching prosperity that are connected to this card, in which the demands of these prosperous characters are very high. The energy is associated with the pressure of being in the spotlight and being made responsible for major decisions that effects those looking on to these characters for guidance. In the BLK Excellence deck, this card is depicted by legendary Hip Hop characters like *Tupac, Biggie Smalls, KRS One, DJ Kool Herc and Doug E. Fresh.*

"More money, more problems." – Biggie Smalls.

In one way or another, these characters became the podium for others to aspire to and consequently needed to maintain an image to appease the fans. This is not to say that they did not believe in the movements that they created. This is to say that they all began their careers in Hip Hop because they were passionate about doing Hip Hop. Out of their passion they were placed in high places and became even more obligated to maintain a specific dynamic. In more recent examples, music performers like Cardi B, A$AP Rocky, T.I., Lil' Wayne and so many others have been put in front of the camera to speak on the Black Lives Matter movement, as a means to get them to lead all black Americans out of oppression. In response to these interviews, most of those artists were baffled as to why they were being asked these questions. As A$AP Rocky stated, "I couldn't relate." In turn the fans were disappointed that these artists were not passionate about activism and political speaking. This is a double edge sword of fame. On one hand performers like Lil' Wayne has been famous since he was 13 years old. So, it is correct for him to say that he cannot relate to living his entire life in the ghetto and being harassed by police. His reality is associated with paparazzi, drugs, parties and yes-men. The same goes for A$AP Rocky and Cardi B. Like the legends of Hip Hop, these performers also followed their passion for money and fame. Therefore, we should ask the question: *is it their responsibility to represent black people in all fronts, regardless if they understand these situations?*

Unfortunately, the answer is yes. Whether they want the responsibility or not, it is the inevitable consequence of fame. You cannot have one without the other.

Advice: The more successful you become, the more responsibility you will have. If you strive to be in the spotlight, then take the initiative to study the responsibilities that you will inherit from your success. Do not try to only focus on the rewards or that position of success will be short lived.

- **Positive Characteristics**: Possession, benefit, inheritance and new awareness.
- **Negative Characteristics**: high-demand, meanness and hasty.

Five of Coins

The energy associated around this card represents flattery and false promises. There are many cases in which we are drawn into a dream that is being presented by a stranger. However, in some cases these promises are only short-lived gratification in which the stranger has no desire to follow through on their word. Everything that glitters, isn't gold. The characters depicted in the BLK Excellence deck shows politician puppets who are collecting their bribe while standing on the stage being held up by starving children. In many cases our world is unfortunately formed off these empty promises that were spoken only for personal gain.

Advice: Be wary of sweet talkers trying to sell you a pipe dream of a better tomorrow. Trust in yourself and your own ability to pool in your resources to build a sustainable future.
- **Positive Characteristics**: Visionary.
- **Negative Characteristics**: Poverty, expenses, loss and struggle.

Six of Coins

The energy associated with this card represents the "act of kindness for a hidden cost." The card is traditionally depicted as a person in a high position giving or helping others, but with a hidden motive. In some cases, this character practices giving for an exchange of sexual favors. In others it could be seen as the energy of a character who is helping a struggling artist, only to later be dictating what the artist creates. As the artist becomes famous, this benefactor becomes the center point in control of the artist's career. The energy surrounding this card is associated with both the manipulator and the manipulated.

Advice: Avoid selling off your soul in hopes of obtaining a little bit of a reward. If you have the skill of refining the skills and talents of others, move to do so without the perversion of common interest.
- **Positive Characteristics**:
- **Negative Characteristics**:

Seven of Coins

The energy surrounding this card is associated with a worker who is devoted to their line of work. Tirelessly they work every day in and day out, working towards next year's harvest. The energy is that of the steady pace building the foundations of a great reward.

Advice: Slow and steady work wins the race. Master your time management skills to ensure that the results of your hard work are successful.
- **Positive Characteristics**: Hard work, success, time management, balance, ambition, abundance, money, growth and joy.
- **Negative Characteristics**: none.

Eight of Coins

Nicknamed the card of Works because it depicts the artisan handling different commissions for their skill. The artisan has become very popular because of their unique abilities, that it has become difficult for them to maintain the demand. Because of this, the artisan has a high standard of which jobs she selects. As she only accepts commissions for projects, she deems worthy for her skill. Even though she is very selective about which tasks she accepts, she is also a workaholic. She must do everything herself and struggles with delegating her tasks off to potential apprentices. Her work and life balance are thrown off, while she becomes more consumed with the quality of her work.

Advice: Perfectionism is a double-edged sword. On one hand, your skill is highly desired and requested. On the other hand, you are so busy that you have no time for friends, family, loved ones and yourself. Let others support you in your work and avoid accepting too much that you can handle because of personal pride.
- **Positive Characteristics**: Beauty, work, skill, mastery, and success.
- **Negative Characteristics**: workaholic, overwhelming and imbalance.

Nine of Coins

The energy surrounding this card is associated with a person who is prosperous or financially secure to live life balanced and comfortably. This person has the energy of an entrepreneur who manages their own time, assets and money. This entrepreneur found success through capitalizing off a significant event. This character is able to hold onto their gains and profits against all the odds. Though looking at their success now, many will overlook the struggles this character had to endure in order to reach this point. This feeling of comfort has been well earned.

Advice: You are in the right to enjoy the comforts of life, as you have earned every moment of it. While you are enjoying these earned moments, find ways to share it with those that supported you with getting here.

- **Positive Characteristics**: Result, comfort, reflection and wealth.
- **Negative Characteristics**: aftermath and realization.

Ten of Coins

The energy that surrounds this card is associated with the energy of generational struggle that has resulted in prosperity with many. One aspect can be found in the sharing of lineage and culture. This prosperity is so abundant that it changes and empowers the destiny of many generations to come.

Advice: You are a benefactor of this reward. Your power to reshape the work to your dreams, like the King of Coins, will not only enhance your life but the life of others. Recognize your true potential and use it.

- **Positive Characteristics**: Black energy, reward, inheritance, influence and satisfaction.

- **Negative Characteristics**: Denial.

Spreads

Knowing the characteristics of the card is one thing, but using that knowledge, is another. Here are some tarot reading layouts to help you speak to your Eternal Self.

Ausar's Spread

The **Ausar** Three Kings spread, mimics the star arrangement for this particular Elder God. Each drawn card represents a stage, related to your question:

(1): **PAST:** the causes that created the situation.

(2): **PRESENT:** the current circumstance.

(3): **FUTURE:** the potential outcome.

Orisha's Spread

The **Orisha** spread is a more of an in-depth answer to a question. Eight cards to match the 8 elder deities:

(1**)**: **Past** - the causes of situation

(2): **Present** - primary issue

(3): **Future** based upon current energy

(4): **Method** to alter frequency

(5): **Current** mentality generating current dilemma

(6): **Additional** energy affecting you

(7): **Challenge** or Barrier

(8): **Potential Outcome**

DOGON SPREAD

<u>*Dogon Spread*</u>

The **Dogon** spread, mimics the star map on the Dogon's traditional masks. This spread maps out the full scenario in question:

(1): **Present**

(2): **Obstacle**

(3): **Intuitive Influence:** energies influencing the situation

(4): **Past energies** effecting you now

(5): **Objective**

(6): **Potential Future** if nothing changes

(7): **Your Mentality**

(8): **Environment**

(9): **Advice**

(10): **Potential Result**: if you follow the advice

Tiye Spread

The **Tiye** spread is to help to you regain balance and harmony with yourself, so that you may become prosperous:

(1): Your skills, talents and gifts that you are aware of.

(2): How you feel about 'said' skills.

(3): Obstacles blocking potential

(4): Revealing the hidden talents you don't see

(5): Advice

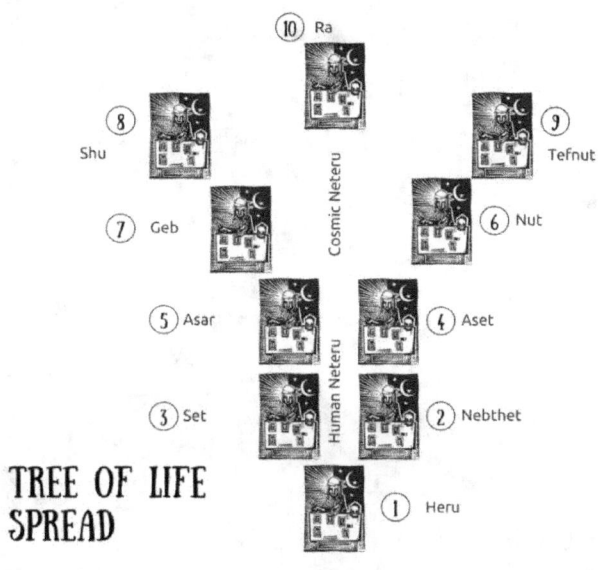

TREE OF LIFE SPREAD

Tree of Life Spread

The **Tree of Life s**pread, outlines our spiritual map to ascending. This is designed to help us keep track of our current spiritual status.

(1): Caution (Pay Attention to)

(2): Quality of Skill

(3): Negative Energy that should be cleared

(4): Obstacle

(5): Overlooked Opportunity

(6): Current Mindset

(7): Current Passion

(8): Advice

(9): Your Strength

(10): Potential Result

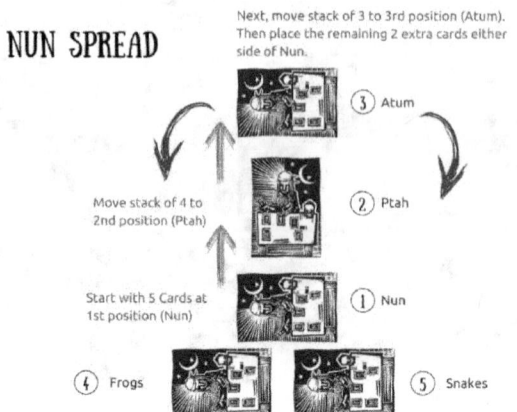

NUN Spread

The **NUN** Spread, outlines the transformation of our desires by mimicking the KEMETIC creation story.

(1): Current State of Mind

(2): The Energy Needed to Progress

(3): The Result of Action

(4): Potential Negative Changes

(5): Potential Positive Changes

Final Response

Throughout this book, we have gone on a beautiful journey.

We have explored the Excellence of Blackness in its many forms. We have learned how to identify toxic imbalanced energies that are oppressive against individualism. We discovered that we are connected to black energy and have access to infinite power dwelling inside of each of us.

We have learned that we are capable of making a change and capable of inspiring others to make a change. In which: *The smallest pebble will sometimes make the largest ripples of change.*

We have learned how to reconnect with our Eternal Selves and have learned how to create powerful communication networks that will help us overcome any obstacle. Finally, we have recognized our potential. With all of these great steps, the only thing that is left for us to do is to put this new understanding into something constructive so that our dreams will become our reality.

In a **Great State of Mind**, we are able to adapt into many **States of Being**. This makes us capable of ascending into a **Final Formation**, that is animated by our **Melanin**. But not just the melanin found within **Us**. We are animated by the melanin of the **Universe...** which is **Black Energy** and **Black Matter**. This is by far, the truest form of **Black Power**. We should therefore strive to share this understanding with every force within our **Existence** and with **Resistance** against those who try to oppress us. This way our actions will speak volumes of our **intellect**, as we allow countless of generations to **Resurrect**... to a maximum. At the **Highest Volume of Frequency**, we will be able to reconstruct this **Reality**. And consequently, will form a better version of you and **me**.

Glossary

3rd Eye Chakra: this is the westernized term for the chakra point located in the middle of the forehead. This chakra point is the key area of the chakra that releases energy that enhance awareness and intuition. This energy point is blocked by Illusion. In Sanskrit it is called *Ajna*.

Afrika: this spelling of the name Africa, is pretty common within the Black Conscious community. It became popular when it was discovered that the letter "C" was developed under ancient colonial powers, such as the Romans. The use of "k" instead of "c" has been a practiced used to separate the conscious community from sources that may otherwise be seen as an exploiter.

Animism: the belief system in which the individual views all living beings (animals, plants, etc.) as having their own consciousness and psyche.

Assiah: this is a Kabbalah Tree of Life term that is used to describe the physical reality.

Asé: the African (Yoruba) translation of the term Chi or Chakra.

ASR: the traditional spelling of the Kemetic deity named Ausar

AST: the traditional spelling of the Kemetic deity named Auset

Aye: the African (Yoruba) translation of the name Earth or the physical reality.

Briah: is a term used in the Kabbalah Tree of Life, which represents the realm of emotional energy.

Candombe: is a religion that formulated in South America (particularly Brazil). This religion is a combination of African deities (particularly the Orishas) and the beliefs of the native South Americans.

Chakra: This is an Indian term which is used to describe the various points in our body where energy is gathered. There are more than twelve chakra points, but only seven points are the most commonly recognized.

Collective Conscious: a pool of energy and information and chain of thoughts that is shared with all living beings throughout reality. This pool of energy influences the actions and behaviors of the beings linked to it. Thus, a fluctuation of evolution that may be spurred by one, immediately influences others to follow or to unconsciously adjust their actions.

Crown Chakra: This is the energy point located at the center and very top of your head. This point is closet (in this dimension) point to reaching the realm of Divine energy. When this energy point is stimulated, it can draw in some divine energy. This point is blocked from Egotistical Attachments. The Sanskrit name is *Sahasrara*.

Cultural Appropriation: is a term used to describe the stealing or taking some aspects of one culture (without the permission of the culture's owners) by members of another culture. This practice is commonly done in western events, such as Halloween, where various groups of other culture's trying to imitate cultures that they themselves do not belong to. For example, a person traditionally associated with Chinese culture, suddenly decides to conduct black facing and do poor impersonations of what they perceive is "black" (which is typically based on stereotypes). Another example would be the five-dollar Indian practices, in which rich or wealthy European (white) Americans would pay to be registered as Indian or Native American. After paying (traditionally five dollars), they would continue to reenact the stereotype perception of what they deemed to be Indian or Native American. As a result, the gene pool of the Native North American, is very inaccurate leaving many perplexed as to how Native Americans look like the colonists who murdered so many of them. It could also be argued that the act of bleaching the skin, bleaching the hair and wearing blue contacts, is also an act of cultural appropriation that African people actively participate in.

Earth Chakra: This is the energy point at your feet that is used for grounding while you mediate.

Elder God: An Elder God is a term that refers to the key elements used in Creation. These key elements are Earth, Air, Water and Fire. Elder Gods are found in most creation stories. In the bible the Elder Gods are described in the seven-day creation story. Outside of religion, these elements are described in western science's big bang theory, too.

Epic of Atra-Hasis: this is an Akkadian tablet that is believed to be the source of the creation story and the Great Flood account that is found in Genesis.

Eternal Self: The Eternal Self is a term that I choose to use in this book, instead of the term "inner child." I found that the term inner child, seemed like a misrepresentation of a being that is both older and wiser than our physical self. The Eternal Self, if you believe in being reborn, has been with you throughout all of your lives. The Eternal Self is your own personal search engine that sees everything and knows everything and who is aware of the consequences of your actions, unlike a child who is carefree and naïve. By identifying that your core is eternal and infinite, you free yourself from the constraints of time. And allow yourself to see passed your own reflection. Yes, we want to see the world through the eyes of a child, without regret. But we also want to be aware of the consequences of our actions. The characteristic is associated with the element of fire, and for this reason, it could be seen as child-like due to the irrational behavior of the element. But fire, as irrational and impulsive as it is, it is also the source of life and the energy within our souls.

Genesis: the cosmology of Abrahamic doctrines (*Christianity, Judaism* and *Islamic*) which tells the account of how humanity was created (sourced from the Epic of Atra-Hasis). Genesis also tells the story of how the world was almost destroyed, leaving the recreation of humans to be reproduced from only a handful of surviving humans (similar to the story of *Ragnarok*). According to the Women's bible (NKJV), the book of Genesis had various different authors. This was determined by researchers, due to the inconsistency of the story and how the style of writing changes throughout the book.

HWT HR: This is the traditional spelling for the Kemetic deity named Hathor.

MTW NTR: This is the name of Kemetic hieroglyphs

NTR: this is the traditional spelling for the name of the Kemetic deities: Neteru

Khememu: the (black) people of KEMETIC

KEMETIC: means black (as in the color black) or the original name of Egypt (Kemet)

KEMETIC NU: this means Black Nation

KEMETIC TA: this means Black Lands

KEMETIC PW: this is a statement which reads "It is Black"

NN KEMETIC PW: this is a state which reads "it is not black"

Qi: Also known as Chi, is a Chinese term used to describe the living energy or life force found in the body. Qi is the main principle behind various Chinese practices, such as Chinese Martial Arts and Chinese Medicine.

Rune Stones: According to Wikipedia "A **runestone** is typically a raised **stone** with a **runic** inscription, but the term can also be applied to inscriptions on boulders and on bedrock." These inscriptions are symbols (which are. In some cases, referred to as letters).

ST KEMETIC: means black woman

Yetzirah: This is the third of the four realms in Kabbalah Tree of Life. This realm is seen as a World of Formation or the element of Air.

You can make a birth chart or natal chart using your tarot cards and your chart. To make one, you can visit any online website to produce your Natal Chart reading. Then, based off your reading, place the cards in the appropriate regions of the star map with astrology symbols. A Tarot Natal Chart is a good way to understand your characteristics, which you can use to further empower yourself.

Every culture has some form of counting system. The Kemetic people also had a counting system, which is very similar to Roman numerals (or rather Roman numerals are very similar to Medtu Neter).

Roughly 3,500 BCE, is when the Cross symbol first appeared, in the Babylonian Empire. The Iron

Cross was the symbol of the Sun God: **Shamash.** There are other variations of the cross that is important to study, which are:

Brighids Cross: Corn dollies are constructed in the shape of Brighids arms pointing to the cardinal points of the compass. The symbol is reminiscent of the ancient Sun symbol, the swastika.

Lorraine Cross: Used to denote one of the degrees within Freemasonry. During WWII, it was adopted by the French Resistance as their secret symbol which stood in opposition to the swastika. The meaning is the Sun.

Cruciger Globus: The orb symbolizes the Earth, and the Cross symbolizes Christ's dominion over it. Or to put it differently, it symbolizes the Sun's dominion over the Earth.

Leviathan Cross: This Cross was used by both the Knights Templar and the Cathars. This symbol became recognized in unorthodox paganism. Similar to the Cruciger Globus, this version of the cross symbolizes the Sun's dominion over the Earth.

Inverted Cross: The "upside down" crucifix, or Cross of St. Peter, is a symbol belonging to Satanists. It is most popularly considered to represent the Anti-Christ, however, the cross being upside down (in light of the style of the Leviathan and Cruciger Globus crosses) could represent the Sun in the planet. The would support the Nazi symbol of the Black Sun in the center of the planet.

Appendix

Zodiac & Christianity

Decoding the 31 Bible Verses about the Twelve Disciples/Apostles

As we have learned, religious stories tend to be stories that are embedded with fact. Or in other words, folklore mixed in with truth, to make it easier to teach a subject. The irony is that many of us (who grew up in modern religious families) were taught that Astronomy was "evil." But as we dive deeper into biblical scriptures, we discover that the scriptures are filled with Astronomical references, which acts as a verbal star map. The following are 31 bible verses that speak about the twelve disciples. The following scriptures are sourced from the King James Version (KJV). The goal of this section, it to teach you how to decode information, so that you can be able to understand the information:

Book of Matthew

Matthew 10:1 Jesus **summoned His twelve disciples and gave them authority** over unclean spirits, to cast them out, and to heal every kind of disease and every kind of sickness.

Note: it is important to notice the line "summoned his twelve disciples and gave them authority" – It could be safe for us to assume, that these twelve that were summoned, are the twelve Houses that surround the Sun. The "summoned" is probably a reference to how the planet orbits into these Houses, giving these

Houses dominion over the behavior and actions of the creatures on the planet.

Matthew 10:2-5 Now ***the names of the twelve apostles*** are these: The first, Simon, who is called Peter, and Andrew his brother; and James the son of Zebedee, and John his brother; Philip and Bartholomew; Thomas and Matthew the tax collector; James the son of Alphaeus, and Thaddaeus; Simon the Canaanite, and Judas Iscariot, who also betrayed him. These [are the] twelve Jesus sent out after instructing them: "Do not go in the way of the Gentiles, and do not enter any city of the Samaritans;

Note: the names of these disciples are Greek. Since we now understand that these are the twelve decans of Kemet (according to Clagett), we can assume that the names of these decans are different, then the names written in the bible. They would also represent the twelve months of the year; or twelve decades (according to Clagett) of the year. Key thing to take note of, is the additional descriptions next to each name, such as "Simon the Canaanite"; which is referring to Simon (aka Peter) as a foreigner to the rest. These minor descriptions tell us more about the character of the name, than the name itself.

- Peter aka Simon = **January** (Aquarius)
- Judas Iscariot = **February** (Pisces)
- Andrew (Peter's brother) = **March**, *when the Sun crosses the equator*
- Matthew = **April**

- John = **May** (who Jesus loved)
- Thomas or Didymus = **June**, the month of June puts an end to the sun's further ascension and begins to shorten all days.
- James (the Great) = **July**
- Judas (brother of James) = **August**
- James Oblia (the Just) = **September**, Libra or Ma'at which is known as the balancer.
- Nathaniel (who Jesus saw under a fig tree, who was gathering the remaining fruits [harvest] of the year. Nathaniel called Philip) = **October**
- Philip = **November** (his name literally means, lover of a horse – Sagittarius)
- Simon, the Canaanite = **December**

Matthew 11:1 *When Jesus had finished giving instructions to His twelve disciples, He departed from there to teach and preach in their cities.*

Note: The Sun, after giving intrusions or authority to the Houses, departs to go teach and preach in their cities. When the planet leaves these Houses, the Sun continues to affect us, even when the Houses (or disciples) have been left behind.

Matthew 19:28 And Jesus said to them, "Truly I say to you, *that you who have followed Me, in the regeneration when the Son of Man will sit on*

His glorious throne, you also shall sit upon twelve thrones, judging the twelve tribes of Israel..."

Note: "**that you who have followed me**" or in other words, the houses surround the Sun, wherever the Sun goes. "**...in the regeneration**" or in other words, in the rebirth (or Spring). "**...when the Son of Man will sit on his glorious throne**" or in other words, when the Sun will sit at its highest peak in the year. "**...you shall sit upon the twelve thrones, judging the twelve tribes of Israel**" or in other words, the Houses or Decans will be a part of the twelve foundations of humanity. Based off our understanding about the influence the celestial bodies have on human awareness and intelligence, we can assume that the twelve foundations of humanity, are the foundations that are key factors to developing human intellect.

Matthew 20:17 As Jesus was about to go up to Jerusalem, He took the twelve disciples aside by themselves, and on the way, He said to them;

Note: "*As Jesus was about to go up to Jerusalem*" could be decoded **as the Sun was about to rise in the East**; as Jerusalem is East of Kemet. "He took the twelve disciples aside by themselves, and on the way, He said to them..." this statement could represent the Sun before it rises on each House, the Sun would meet with them, one on one. Meaning, from our perspective, before we would see the Sun rise for each Zodiac period, the Sun would be at the House of this sign before it rises. On a scientific bases, this is how it works. Well before we can see the Sun rising in a new astronomical house, the Sun meets at the entrance of the House, as though it was speaking with each "disciple" before it entered their House.

Matthew 26:14 Then one of the twelve, named Judas Iscariot, went to the chief priests

Note: Judas the Betrayer, is the characteristic of the House of Scorpio/Wadjet. Wadjet is considered the Protector (Goddess) of the Lower Lands of Kemet; which is a region known to constantly be at war, throughout history, with Upper Kemet. Therefore, it could be assumed that this character (or region) would go to their chiefs to betray their lords.

Matthew 26:20 Now when *evening came, Jesus was reclining at the table with the twelve* disciples.

Note: it is important to pay attention to the timeframe of this scripture *"when evening came…"* What happens in the sky, in the evening? The Sun begins to set or *"reclines at the table"* or at the horizon, with the twelve Houses.

Matthew 26:47 While He was still speaking, behold, Judas, one of the twelve, came up accompanied by a large crowd with swords and clubs, who came from the chief priests and elders of the people.

Book of Luke

Luke 6:13 And when day came, He called His disciples to Him and chose twelve of them, whom He also named as apostles:

Note: the breakdown of this line would be:

"When the day came..." = when the Sun rose...

"He called His disciples to Him" = the Sun shown on the celestial Houses around it

"chose twelve of them..." = the twelve houses surrounding the Sun

"He also named as apostles" = the word **apostle** means "messenger." This is important to take note of because in other scriptures that reference to the disciples, the Sun, gives these Houses *authority over mankind* or the power to influence our behavior and intellect. Thus, it would make sense as to why, these Houses are called *messengers*.

Luke 8:1 Soon afterwards, He began going around from one city and village to another, proclaiming and preaching the kingdom of God. The twelve were with Him,

Luke 9:1 And He called the twelve together and gave them power and authority over all the demons and to heal diseases.

Luke 9:12 Now the day was ending, and the twelve came and said to Him, "Send the crowd away, that they may go into the surrounding villages and countryside and find lodging and get something to eat; for here we are in a desolate place."

Luke 18:31 Then He took the twelve aside and said to them, "Behold, we are going up to Jerusalem, and all things which are written through the prophets about the Son of Man will be accomplished.

Luke 22:3 And Satan entered into Judas who was called Iscariot, belonging to the number of the twelve.

Luke 22:47 While He was still speaking, behold, a crowd came, and the one called Judas, one of the twelve, was preceding them; and he approached Jesus to kiss Him.

Book of Mark

Mark 3:14 And He appointed twelve, so that they would be with Him and that He could send them out to preach,

Mark 14:43 Immediately while He was still speaking, Judas, one of the twelve, came up accompanied by a crowd with swords and clubs, who were from the chief priests and the scribes and the elders.

Mark 4:10 As soon as He was alone, His followers, along with the twelve, began asking Him about the parables.

Mark 6:7 And He summoned the twelve and began to send them out in pairs, and gave them authority over the unclean spirits;

Mark 9:35 Sitting down, He called the twelve and said to them, "If anyone wants to be first, he shall be last of all and servant of all."

Mark 10:32 They were on the road going up to Jerusalem, and Jesus was walking on ahead of them; and they were amazed, and those who followed were fearful. And again He took the twelve aside and began to tell them what was going to happen to Him,

Mark 11:11 Jesus entered Jerusalem and came into the temple; and after looking around at everything, He left for Bethany with the twelve, since it was already late.

Mark 14:17 When it was evening, He came with the twelve.

<u>Book of Revelations</u>

Revelation 21:14 And the wall of the city had twelve foundation stones, and on them were the twelve names of the twelve apostles of the Lamb.

<u>Book of John</u>

John 6:70 Jesus answered them, "Did I Myself not choose you, the twelve, and yet one of you is a devil?"

John 20:24 But Thomas, one of the twelve, called Didymus, was not with them when Jesus came.

John 6:67 So Jesus said to the twelve, "You do not want to go away also, do you?"

Book of Acts

Acts 6:2 So the twelve summoned the congregation of the disciples and said, "It is not desirable for us to neglect the word of God in order to serve tables.

Acts 19:7 So the twelve summoned the congregation of the disciples and said, "It is not desirable for us to neglect the word of God in order to serve tables.

Book of First Corinthians

1 Corinthians 15:5 and that He appeared to Cephas, then to the twelve.

Zodiac Statements

The Zodiac houses are determined by their celestial placement on the map. Throughout the year, as our planet orbits around the Sun, we move into different houses. As we orbit through each of these houses the development and actions of the creatures on the planet, are affected. As we are born, we are embedded with the characteristics of the house and the decan, we were born into. In other words, the energies that exist on the planet due to our planet's placement in the solar system have an impact on the character and personality development of individuals being born. These characteristics are best summarized by their Zodiac Motto, which are:

AUSAR (Aries) 21 March to 19 April | Statement: "I am"

AUSET (Taurus) 20 April to 20 May | Statement: "I have"

AUSAR (Gemini) 21 May to 20 June | Statement: "I think"

MA'AT (Cancer) 21 June to 22 July | Statement: "I feel"

SEKHMET (Leo) 23 July to 22 August | Statement: "I will"

TEHUTI/DEJUTI (Virgo) 23 August to 22 September | Statement: "I analyze"

KHUNUM (Libra) 23 September to 22 October |
Statement: "I balance"

WADJET (Scorpius) 23 October to 21 November |
Statement: "I lust"

NWT (Sagittarius) 22 November to 21 December |
Statement: "I see"

HERU (Capricorn) 22 December to 19 January |
Statement: "I use"

PTAH (Aquarius) 20 January to 18 February |
Statement: "I know"

KHEPERA (Pisces) 19 February to 20 March |
Statement: "I believe"

The Concept of Kemetic Creation Story

There is a famous cosmology story that was found on an ancient carving. This carving is called the Shabaka Stone, which is summarized as:

First, there were the waters of NUN who we call the Divine. For the Divine to exist, it needed an identity. Therefore, the ENERGY at rest turns into ENERGY in motion, which causes PTAH (a hill) to rise out of the waters of NUN. Once, PTAH rose up ATUM separated itself from PTAH, then attached itself on top of PTAH. This made the ATUM the unmoved mover. From ATUM, four pairs of beings were created, so that NUN may be given a name and an identity. These four pairs of beings remained in the waters of NUN as frogs and snakes until it was time for them to transform (KHEPRA)

The four pairs of beings were:

- **Energy of Motion/ Energy at Rest**
- **Infinite and Finite**
- **Dark and Light**
- **Revealed and Hidden**

Throughout African Spirituality, you will find a similar story of supreme energy which creates four pairs of deities (each pair consists of masculine and feminine... the balance) It was during and after colonialization when this concept of a balance of masculine and feminine energies was changed to be primarily masculine. This abundance of masculine energy in our society is the primary reason we are out of balance. The only way we can repair this damage is by reconnecting to the Divine or ourselves. We each hold a piece of the Divine within our being, this is the energy that flows through us which is called Asé. Asé flows through our bodies and is the energy that helps us connect to the energy of the universe as well as the energy that helps us alter our reality. That is why Asé is typically said after every statement or prayer; as Asé is the will to make all that was spoken to manifest into this reality. As Africa was invaded or as we expanded out of Africa, many of the practices were used in other parts of the world in other cultures with new names.

Kemetic Deities

Ausar (ASR):

Ausar is usually shown wrapped partially as a mummy, except for the upper part of his body with his arms emerged and holding a flail and a crook (scepters of kingship). He wears an Atef Crown. He represents renewal, which is why only half of his body is wrapped for death and the other half is left for life. His story is that he was broken into many pieces and put back together again, making him the NTR of new life.

Element: Fire

Auset (AST):

Auset is the original mother Mary, who holds her child, Heru. She is usually sitting on the throne with her son on her lap and with a crown on top of her head. She symbolizes motherhood, intuition, nurturing, empathy and fertility. She is also a protector and her energy are to bring us completion and awareness.

Element: Water

Tehuti:

The language MDW NTR is said to have been created by him, who bestowed it upon our ancestors. His powers are related to recording of facts and of data associated with the collective conscious and the Divine. He uncovers the lost knowledges but can easily communicate with the mind via the 3rd Eye. His power of communication is so great that he can calm the chaotic nature of mankind with a whisper.

Element: Fire

Ma'at:

She is usually depicted with her arms stretched out (wings) and in a kneeling position with a feather on top of her head. The single feather is also the corresponding MDW NTR symbol for her name. Our ancestors spoke her laws to ensure that they could ascend and pass the judgement, as it was her laws that taught us the importance of Balance.

Element: Air

Shu:

Displayed in a male form, who is most commonly known for separating the earth (Geb) from the sky (Nut). The key symbol for Shu is the feather on his head, which represents air. He is the oxygen that we breathe in. Because he is the symbol of Air, he can fit into any situation and bend around any obstacle.

Element: Air

Ptah:

He is the symbol of moving energy or animation. He was created by moving from the unmoved

(Nun) and from Ptah, Atum separated itself then reattached itself on top of him. This can suggest that from Ptah going into action, he then is the cause or the influence of others going into action.

Element: Earth

Sekhmet:

She is referred to the Eye of Ra or the Powerful One. She is also the source of the saying "never use his name in vain" as she was such a sacred NTR that the Kemetic people didn't utter her name. It was believed that by speaking her name, that those ones would be destroyed. She is the protector of warriors as well as a protector of healers. She is the goddess of war and the destroyer of the enemies of Ra, her father.

Element: Fire

Khunum:

He is usually depicted with the head of a ram and he represents the energies that bring excitement to our existence. He symbolizes the energy of Inspiration.

Element: Earth

Tarot Natal Chart

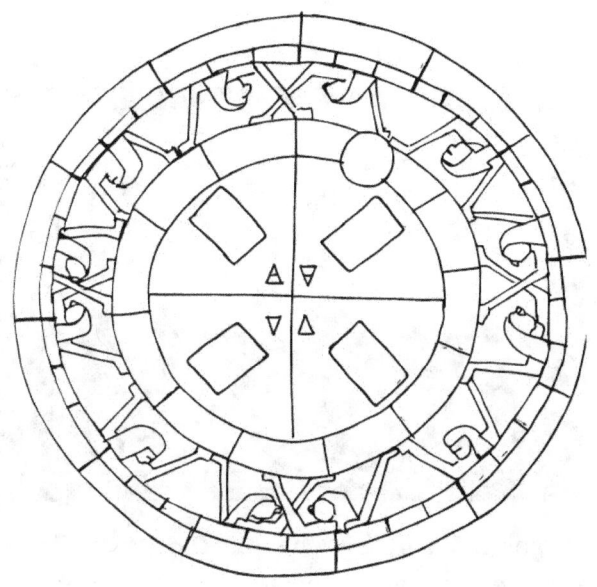

You can make a birth chart or natal chart using your tarot cards and your chart. To make one, you can visit any online website to produce your Natal Chart reading. Then, based off your reading, place the cards in the appropriate regions of the star map with astrology symbols. A Tarot Natal Chart is a good way to understand your characteristics, which you can use to further empower yourself.

Evolution of the Cross

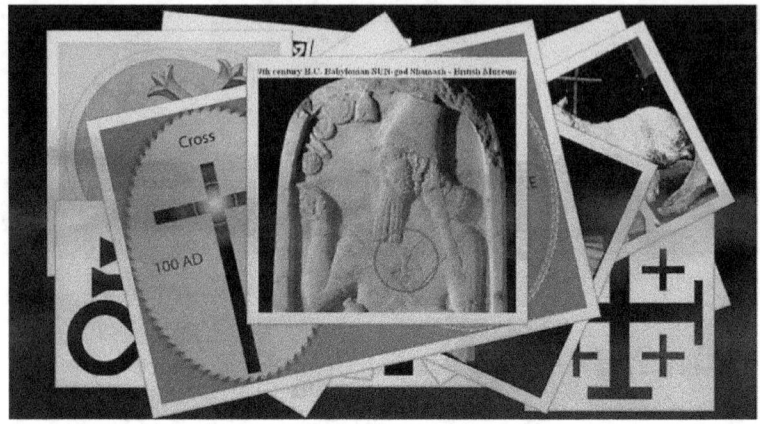

Roughly 3,500 BCE, is when the Cross symbol first appeared, in the Babylonian Empire. The Iron Cross was the symbol of the Sun God: **Shamash.** There are other variations of the cross that is important to study, which are:

Brighids Cross: Corn dollies are constructed in the shape of Brighids arms pointing to the cardinal points of the compass. The symbol is reminiscent of the ancient Sun symbol, the swastika.

Lorraine Cross: Used to denote one of the degrees within Freemasonry. During WWII, it was adopted by the French Resistance as their secret symbol which stood in opposition to the swastika. The meaning is the Sun.

Cruciger Globus: The orb symbolizes the Earth, and the Cross symbolizes Christ's dominion over it. Or to put it differently, it symbolizes the Sun's dominion over the Earth.

Leviathan Cross: This Cross was used by both the Knights Templar and the Cathars. This symbol became recognized in unorthodox paganism. Similar to the Cruciger Globus, this version of the cross symbolizes the Sun's dominion over the Earth.

Inverted Cross: The "upside-down" crucifix, or Cross of St. Peter, is a symbol belonging to Satanists. It is most popularly considered to represent the Anti-Christ, however, the cross is upside down (in light of the style of the Leviathan and Cruciger Globus crosses) could represent the Sun on the planet. The would support the Nazi symbol of the Black Sun in the center of the planet.

References

Amos N. Wilson | *Special Education: Its Special Agenda Unhooded* [Video file]. (2015, May 25). Retrieved from https://www.youtube.com/watch?v=3QQA6ObBcRo

Ancient African Mathematics. (n.d.). Retrieved from http://www.taneter.org/math.html

Ani, M. (1994). *Yurugu: An African-centered Critique of European Cultural Thought and Behavior* (1st ed.). Trenton, NJ: Africa World Press, Inc.

Ashby, M. (2007). *The Kemetic Tree of Life: Ancient Egyptian Metaphysics and Cosmology for Higher Consciousness* (1st ed.).

Assaan-Anu, H. Y. (2010). *Grasping the Root of Divine Power*

Bell, H. J. (1893). *Obeah: Witchcraft in the West Indies.* London, United Kingdom: Forgotten Books.

Binary Numbers in Ancient India. (n.d.). Retrieved from http://home.ica.net/~roymanju/Binary.htm

Binary number. (2003, June 2). Retrieved from https://en.wikipedia.org/wiki/Binary_number

Blake, H.(2009) 'Michael Jackson's history of heath problems'. Telegraph.[Online]26 June. Available at http://www.telegraph.co.uk/culture/music/michael-jackson/5650170/Michael-Jacksons-history-of-health-problems.html

Bobrow, RS.(2011) *Evidence for a Communal Consciousness.* Unpublished Dissertation. NCBI /

Bynum, E. B. (2012). *Dark Light Consciousness: Melanin, Serpent Power, and the Luminous Matrix of Reality*

Crick, F. C., & Koch, C. (2005). What is the

function of the claustrum? *Philosophical transactions of the Royal Society of London. Series B, Biological sciences*, 360 (1458), 1271–1279. doi:10.1098/rstb.2005.166

Cooper, A.(2010) 'Study: White and black children biases toward lighter skin'. CNN.[Online] May 14, 2010.Available at http://www.cnn.com/2010/US/05/13/doll.study/index.html,[Accessed February 3 2019].

Drake, J. (2000). Native American Wars. In the Oxford Companion to American Military History. : Oxford University Press. Retrieved 12 Sep. 2019, from https://www.oxfordreference.com/view/10.1093/acref/9780195071986.001.0001/acref-9780195071986-e-0618.

Dr Ben - *Belief Vs Knowledge* [Video file]. (2015, March 20). Retrieved from https://www.youtube.com/watch?v=N5l2uBUyY0g

Dr Frances Cress Welsing - *Noose, Swastika and Burning Cross* [Video file]. (2014, March 6). Retrieved from https://www.youtube.com/watch?v=AnDPVB60c8g

Dr. Mercola. (2014, August 14). Study: Hereditary Trauma Are Passed on to Children Through Smell. Retrieved from https://articles.mercola.com/sites/articles/archive/2014/08/14/hereditary-trauma.aspx

Flury, M. (2015). *Downloads from the Nine: Recognize Your Higher Self Effortlessly.*

Gaia Staff, the (2017) 'YOU MIGHT HAVE PSYCHIC ABILITIES THANKS TO A COLLECTIVE CONSCIOUSNESS'.GAIA.[Online] July 14.Available at http://https://www.gaia.com/article/you-are-contributing-to-humanitys-collective-consciousness

Garvey, A. J. (2013). *The Philosophy and Opinions of Marcus Garvey: Africa for the Africans*. London, England: Routledge.

459

Hearing Different Frequencies. (2014, June 2). Retrieved from https://www.nih.gov/news-events/nih-research-matters/hearing-different-frequencies

Hilliard, A. G. (1999). *Ancient Africa's Excellence Tradition*. (Speech)

History Disclosure Team, the. (2016) *'Science Provides Strong Evidence on the Existence of a Collective Consciousness'*. History Disclosure.[Online] April 11. Available at http://http://www.historydisclosure.com/collective-consciousness/

Isichei, E. (1995). *A History of Christianity in Africa: From Antiquity to the Present*. Grand Rapids, MI: Wm. B. Eerdmans Publishing.

Ivan Van Sertima – *They Came Before Columbus: The African Presence in Ancient America*. (1976)

Ivan Van Sertima | *Egyptian Science* (10 May 1989) [Video file]. (2015, March 5). Retrieved from https://www.youtube.com/watch?v=l1oZyXFDY7A

John, PhD, Y. (2016, October 21). *Where Do Our Thoughts Come From*? Retrieved from https://www.forbes.com/sites/quora/2016/10/21/where-do-our-thoughts-come-from/

JoyDailyTV. (2016).Cake Soap - Skin Bleaching in Jamaica.[Online Video].December 4 2016.Available at http://www.youtube.com/watch?v=JmY0_l6BNPc

Leonte, A. (2016). *The Alphabet of Desire - a Method of Chaos Magick* (1st ed.)

Massey, G. (2007). *A Book of the Beginnings*. New York, NY

MATHEMATICS OF EGYPT - Mathematicians of the African Diaspora. (n.d.). Retrieved from http://www.math.buffalo.edu/mad/Ancient-Africa/mad_ancient_egypt.html

McLean, B. (n.d.). Legions of Light:

Interview with an Aboriginal Woman. Retrieved from http://www.seizethemagic.com/lol/05worldinfo/05abowom.html

Morris, J. (n.d.). The Power of Vision Boards. Retrieved from https://www.selfgrowth.com/articles/the-power-of-vision-boards

Morsella Ph.D., E. (2012, February 9). *What is Thought*. Retrieved from https://www.psychologytoday.com/us/blog/consciousness-and-the-brain/201202/what-is-thought

NASA (unknown) *Dark Energy, Dark Matter*.[Online] Available at http://www.science.nasa.gov/astrophysics/focus-areas/what-is-dark-energy

National Institute of Health. (2014, June 2). Hearing Different Frequencies. Retrieved from https://www.nih.gov/news-events/nih-research-matters/hearing-different-frequencies

NCADV | National Coalition Against Domestic Violence. (n.d.). Retrieved from https://www.ncadv.org/signs-of-abuse

NCBI / Anna-Kaisa Newheiser and Kristina R. Olson (2012) *White and Black American Children's Implicit Intergroup Bias*.[Online] Available at http://https://www.ncbi.nlm.nih.gov/pmc/articles/PMC3241011/

Noose, Swastika and Burning Cross- Dr Frances Cress Welsing [Video file]. (2014, March 6).

Numbers in ancient Egypt. (2018, April 19). Retrieved from https://quatr.us/egypt/numbers-ancient-egypt.htm

Rajpal, K. L. (2013). Dark Photons and Dark Matter. Retrieved from http://vixra.org/pdf/1303.0207v3.pdf

Ren, T., He, W., & Barr-Gillespie, P. G. (2016, January 6). *Reverse transduction measured in the living cochlea by low-coherence heterodyne interferometry.* Retrieved from https://www.nature.com/articles/ncomms10282

Prehistoric Mathematics. (n.d.). Retrieved from http://www.storyofmathematics.com/prehistoric.html

Prof Emeritus John Henrik Clarke - African World Revolution - Africans at the Crossroads [Video file]. (2016, November 21). Retrieved from https://www.youtube.com/watch?v=p1jvCjk5f4k

Shawna Freshwater, PhD (2017)1st Chakra Root Muladhara.[Online] Available at http://www.spaciousthe rapy.com/1st-chakra-root-muladhara[Accessed February 3 2019]

Siegal, E.(2018) *'This Is How Much Dark Matter Passes Through Your Body Every Second'*. Forbes Magazine.[Online]July 3.Available at https://www.forbes.com/sites/startswithabang/2018/07/03/this-is-how-much-dark-matter-passes-through-your-body-every-second/ ,[Accessed February 17 2019].

Science Daily (2017) *Children show implicit racial bias from a young age, research find*.[Online] Available at http://https://www.sciencedaily.com/releases/2017/11/171127152100.htm [Accessed February 20, 2019]

Sowell, T. (2006). *Black Rednecks and White Liberals*. Encounter Books.

Storr, A. (2015). *Solitude a Return to the Self*. New York, NY: Simon & Schuster.

The Ishango Bone: Cradle of Ancient Mathematics. (2013, September 21). Retrieved from https://afrolegends.com/2013/08/29/the-ishango-bone-craddle-of-mathematics/

Unknown. (n.d.). Tehuti. Retrieved from http://tehuti.org/tehuti.html [Accessed August 16, 2019]

What Are the Rules of Mancala? (n.d.). Retrieved from https://www.reference.com/hobbies-games/rules-mancala-1e943264728b728d#

Williams, E. E. (1964). Capitalism & Slavery

Woodson, C. G., & James, G. G. (2014). *The MIS-Education of the Negro and Stolen Legacy.*

Young, V. (2011). *The Secret Thoughts of Successful Women: Why Capable People Suffer from the Impostor Syndrome and how to Thrive in Spite of it..*